STELLAR SLAPSTICK

ELEANOR MORRIS WU

Published by:

Light Switch Press

PO Box 272847

Fort Collins, CO 80527

Copyright © 2017

ISBN: 978-1-944255-52-7

Printed in the United States of America

PREFACE TO STELLAR SLAPSTICK
BY ELEANOR MORRIS WU

Eleanor Morris Wu was born in Lancaster, PA and graduated BA from Harvard College (English, honors) the cum Laude, as well as intense study in other fields such as mathematics, French, history and Biochemistry.She also graduated MA in Chinese Anthropology from the University of Toronto in 1992. She also is a doctorate from China an Academy in Taiwan 1971 and is a Litt. D [HON] Council of Arts and Culture, UNESCO, LA 2004. She has published 14 books on Amazon.com https://goo.gl/8Rh6y9

You can know more about her works from the below links:

https://goo.gl/8DjPbs

https://goo.gl/yp5PPD

She has written a new Physics book inspired from the Archives of Ancient Chinese, Science and Chinese, I Ching, Stellar Slapstick and Surrealism and Dadaism of deep space and will be exploring end purposes of inverse activity and that will lead to conversation of entropy through use of surrogate in the solar planet systems, planets as differentials of the solar sun as well as meme like planets will be presented.

The issues that are developed in her book are as follows:

- Using ancient Chinese concepts of Yin/Yang and Number theory, they have solved the oldest problems in science brought up by ancient Greeks, squaring of the circle and the wave-particle duality. Theories of ancient Chinese science shown to be validated in Einstein's theory of General Relativity.

- New ways of using ancient Chinese science to get information about DNA.

- She has also discussed other things like the machine that produces perpetual energy of the sun. Origin of Black holes.

- Is there any other place in our solar system that can support human-like life?

- How to tell if exo planet can support human-like life or life at all?

- What is relevance to human life on our solar system to life in the universe as a whole?

In Chapter 10 of Stellar Slapstick,Using the spectroscopic research and the theoretical explanations of Hoyle in the '50's buttressed by more recent research in the new chemical field of element abundance in the universe,Morris Wu contends, the gap between objective science and religious faith has been conclusively closed not by arguments of faith but cold hard science.Morris Wu explains Hoyle's spectroscopy showed that in a secondary process of element production in the fiery furnaces of stellar stars 8-10 times larger than our sun, a definite bias was shown for the production of elements vital for life in planetary systems such as our own as well as for the construction of these planetary systems themselves enabling life forms. When the life of these fiery furnaces was spent, massive residues of life-giving elements were spewed in stellar clouds and winds across the universe landing and coagulating on our planetary system, for one, ultimately creating the earth and all her myriad life forms on it.Morris Wu's conclusion is that It is this bias or choice for life in element creation that Intelligent Design or the Mind of Divine Creator can be scientifically authenticated-- **Dhruv Dave, Ask Sunday Services**

Summary of new planetary science: quest for life in our outer solar system and beyond; terrestrial planets, gas, ice giants, exo planets, astro biology and space junk and other stellar slapstick, surrealism space and deep space Dadaism

By Eleanor b Morris WU

© copyright 2016, Alexandria, VA, USA

Personal preface dedicate to other space travelers on the crowded highway of space junk.

Reading through conversations of Facebook expat friends in Taipei, for the last few months I was rolling in the isles. Snl has nothing on the conversation going on between Crack, Chris Fay, Bruce and McGee. Bruce's comment that Crack will be riding around our galaxy painting 'Killroy was here' on the planets, I mean just so extra-terrestrial, meme minded. Have you guys seen or heard of the funniest comic strip going today 'Brewster rocket space guy' by Tim Richards I wrote to these friends?, veritable stellar slapstick Richards knocks the stuffing out of all popular space icons from Star trek to Lord of the rings, satirizing everyone from Captain Kirk to Yoda and princess Leia. Our real life space whizes don't escape either, in one of the latest strips Richards gives some good hoots on space engineers like Elon Musk who plans to build human colonies on Mars even though Mars has only enough oxygen for a few roaches. In his latest trip idiot Commander Brewster hires some space aliens to renovate earth. They plan replacing out oxygen atmosphere with methane gas, entirely disregarding the first priority for humans on the planet. The price for renovation goes up due to the fact there are too many tectonic plates under earth's surface, upping the price for granite table tops. The alien renovator plans to rearrange the tectonic plates for aesthetic reasons and spackle the Grand Canyon. It goes on, have a read, dear readers as I send my love to you and all my old buddies at the watering holes of Taipei and Bangkok from here in mellow Virginia, in prosperous USA

Yours Ellie ancient Chinese science by Eleanor Morris Wu

I am inspired to write my new mathematical treatise entitled ''manifesto of mathematical memes of the 21st century, après de fin de siècle, featuring popular dance moves and mass movements of our time, for example 'The Trumpian Twist', the 'spoiled kid's shuffle of a grand new era [when daddy trump told baby trump, 'if you believe it's true, it's true'. Baby trump took it literally, leading to featured dances. 'The trump twist; and 'spoiled brats shuffle', leading into one prochaine siècle's the 22nd century, stellar slapstick, space surrealism odyssey and 'deep space Dadaism 0pera

In my new mathematical treatise of the universe I had decided to take the perspective from old Chinese science [and philosophy] and apply it--or reintegrate it-with new findings in our modern western mathematics and science. Using this method I think it would be safe to say that ancient Chinese science was an egalitarian science, a science meant for the common man and woman whereas our modern western science and mathematics is definitely the science of the eggheads, meant for the rich and entitled, those who could afford pricey and prestigious educations in the lived halls of Oxford or Cambridge, or in the US, of Harvard, MIT and Stanford. Thus to get a marketable degree, one that would enable the scions of titled, the wealthy and the corporate elite, to get a six figured salary upon graduation from one of these elite institutions, these fortunate scions would be commanded by their successful progenitors to take up degrees in higher mathematics and advanced physics or financial management. There they would be expected to digest and regurgitate on request lengthy and complicated equations of the partial derivatives of the iconic curve on the radiant tangent of cost flow or nuclear generation [or random manufacturing] to electromagnetic radiation[to inflation thrust]/ the proper depth of understanding of these equations by honored graduates was their total inability to explain the meaning of these equations either in comprehensible written English or verbal comprehension to others. What these equations meant or what they referred to, how they mattered in the real world were answered only with an elitist ivy shrug. The message was always clear though the answer was totally obscure.

Using the influence of ancient Chinese science however parents of mr e modest economic backgrounds and lineage descent arrays might well in --a more egalitarian future--sent their offspring to get degrees in 'lower mathematics and regressive physics--or financial mismanagement. Here the validity and authenticity of their degrees would be measured not but incomprehensibility of the complex equations they had been taught to spout but by the arcane simplicity of the simple formulas easily translated by them into simple English, both written and spoken, easily understood by the least privileged or wealthy of their age peers, those who had never graduated from oxford, Cambridge, MIT or Stanford, but had, in the new egalitarian age, made do with Oxford, Cambridge, kit and ham word, pale

CHAPTER 1

Today as those of us of retrenched Salvationist faith of science fiction soaked adolescence and youth, come into maturity-if not middle or old age-to see the phantasies of storied universes, galaxies and foundations, take on an overview reality as we see or science fictive peers prepare to send ships into our outer solar system and colonize mars with real people. Although some of our more conservative brethren are cautioned but a knowledge of the stark lack of material resources necessary to human life in the outer solar system, the wild enthusiasm of our more progressive peers threaten to drown out any plaintive fears of mass genocide on distant planets, such wild enthusiasm fueled but so many mass genocides in futile religious wars on earth in the preceding few hundred years and even today. To these visionaries mass genocides could not happen to colonists who did not carry any of earth's toxic religions with them, but were good, clean, non-tainted non-believers and atheists, ready to bring good clean healthy science to ;brave new worlds', long tainted, long discredit religious belief systems all based on outdated and irrational superstition and exclusive cults.

Thus while we have seen that western modern science does have its own arcane vocabulary of mostly incomprehensible phases, key words, equations that are reciters obsessively as any believers in modern religious litany, we cannot deny that your basic advances in science in all branches and technologies and engineering which give definition to life in the developed and developing societies on earth are built on the back of modern western science. While the hoop of modern western science is the sole preserve of the scientists-priests-eggheads-elitists are the deacons of its church, an almost infinite variety of useful products are the end result of the science myth. Part hootenanny. Part valuable reality

With all the hoot nay however has come some important problems needed to be solved by science without hoot nay: to what extent ids the near hysterical ranting's of a post science fiction generation real about unknown bounties awaiting us in our outer solar system and beyond and how much of it did just that--hysterical phantasy that will take our next generation like lemmings into the sea as they follow the hollow melodies of egghead pied pipers?

It seems to me the most practical way to approach this problem is by a systematic theory of solar systems and galaxies and our solar system and galaxy in particular. However the feat of developing theory of planetary systems in the west was last tried on middle ages theologian, St Thomas Aquinas who, using Aristotelian logic from the ancient world, combined with hysterical devotional Christian theology, determined that all events, causes and effects in both the real and spiritual world could be determined by counting the angels nesting on the backs of all material phenomena. While St Augustine's in cogent determinism was finally disproved by the equally in cogent 19th century German idealism of, no new 'silver bullet' ideology [not counting egregiously fallible materialism theory of Karl Marx]has been tried again in the western world.

For a more fruitful way of trying to divine the mysteries of creation using some sort of modern western logic, I would opt for the simplest method, that of pure mathematics. But to undo our purest math from the philosophic and religious fogs rife in the western world I will look at our mathematical theories through the purer and cleaner prism of ancient Chinese science and math, at least to the extent to which I am able to understand them.in other words I take my trip understanding the causes and effects of the real/material world on the tram that travels from the higher math and advanced physics of the west to the lower math and regressive physics of ancient china.

————

Summa of new planetary science,: quest for life in our outer solar system and beyond; terrestrial planets, gas, ice giants, exo planets, astro-biology and space junk and other stellar slapstick, surrealism space and deep space Dadaism

Lower mathematics of ancient Chinese science

CHAPTER 2

After the fall of Rome in the last millennium and the chaos that followed everywhere in the western world, science and math were kept alive but vigorous exchanges with scholars in the Arab world and the Indian continent. Westerners learned algebra and eventually alchemy from the Arabs, embarking on a long and largely unproductive road into the medieval world of the west until the Italian renaissarti facts of ancient Chinese science. Ice turned everything around for the west. After Copernicus's daring refutation of papal science a pletors of budding astronomers and navigators coming out of the Mediterranean kingdoms, looking for trade routes to the riches of the orient, invented critical astronomical and navigational tools that jettisoned the west into the technological power, it is today. As accuracy of telescopes and continually improved, astronomers such as Kepler and Tycho graph brought in the modern scientific revolution. This was capped by the work of the great Italian astronomer and mathematician Galileo.

While historians of western science have always said origins of science in the western world have always said the origins of scientific findings in the ancient world were unclear, closer examination of ancient Chinese civilization, which predates ancient western civilization but as many eons, makes it clear that ancient western science was a primary or tangential derivative of ancient Chinese science. The use of the binary, the dialectics, discovery of pi, hypotenuse of right angle triangle and all logic and mathematical rigors of Chinese in Ching and other artifacts of ancient Chinese science.

Even the architectural wonders of the ancient Egyptians with a combination of precise geometry and sorcery, picture language, Shamanistic religion and sorcery finds its roots in rage ancient Chinese civilization. Further remnants of sorcery, magic and wizardry of the ancient Hebrews also finds its roots there.

Even further back into the Paleolithic civilizations such as Stonehenge of ancient Celts and Britain's, preceding even picture writing systems, near the beginning of the first homes appearance on earth, civilizations built entirely on sorcery and wizardry, orient Chinese civilization are to be found

While it is often said that marks modern western science apart from science I any other or any other civilization is the practice of verification to proof and meticulousness of measurement and calculation of experimentation in natural and physical world. Such meticulous calculation and measurement is made possible by western scientists for these purposes.

And while complex measuring instruments that might have use by ancient Chinese scientists have been lost in the sands of fast receding ancient millennium, it would be the utmost hypocrisy to deny that experimentation and meticulous measurement has not been a hallmark of Chinese science since the time Chinese history has first been recorded been recorded more than 5000 years ago. These devices such as the compass have been incorporated in western science since the early western ancient era.

While cardinal numbers have had extreme religious significances in the religions of the ancient western world, the actual deification of cardinal numbers are found in the earliest extant religious texts of china, predating the ancient western world by as many eons. For example in the holy tests of the book of Tao, perhaps china's oldest religion, the nub 1--which is also the symbol of Tao is the ruling deity of all deities in the Tao canon. The remaining cardinal numbers up to single digit number 9 are similarly deified or canonized with and imputed with decreasing power as the size of the number increases.

The significance of the special powers of particular cardinal numbers in Chinese religion cannot be underestimated, for this Chinese 'number theory' in religion was the basis of their ancient science and since their ancient science comprised all branches of science that we in the west recognize today and more, from physic, to chemistry, to meteorology to geology to agricultural science to pharmacology to medicine to psychiatry to psychology to sociology--to astronomy to cosmology to astrology and beyond, it can be seen as having profound and pervasive influence of every phase of life for every Chinese person since time immemorial

I am attempting to integrate basic concepts of ancient Chinese science with modern western science the task has been made wonderfully easier since the discovery of DNA in early 1950'w by Francis Crick of Oxford and James Watson of Harvard. DNA defy nuclei acid is a molecule comprised of 5 of the atomic elements in the periodic table of the elements, with cardinal atomic numbers 1=Hydrogen, 6=Carbon, 7=Nitrogen, 8=Oxygen, 0=Phosphorus linked together on a molecular string making semi closed circle of pentagon shape. DNA has been verified as the all-important, indispensable molecule for all living things on this planet. Where dies DNA come from? Is it only native to earth? But it has been found on comets and asteroid plummeting to earth from outer solar system. And why these5"

Miraculously the DNA molecule was a vital component of ancient Chinese science.

Although ancient Chinese science did not have a periodic table of the elements with quasi scientific terminology naming all the elements its ''5 indispensable elements of life' comprised of 5 natural phenomena most noted for containing appreciable quantities of the five elements of the table described by Crick-Watson.

Thus for the 5 indispensable elements of life for ancient Chinese science and their chemical equivalents in the DNA molecule fire=hydrogen, wood=-carbon, earth=nitrogen, water=oxygen, metal=phosphorus like the semi closed circle of DNA molecule shape, the 5 indispensable elements of life vouched for by ancient Chinese were strung into the shape of a semi closed circle or pentagon.

And just as the modern organic molecule 3 was awash with bonding joints at every intersection of the 5 elements so the pentagon pf the Chinese 5 elements indispensable for life was an a set of almost infinitely potential other pentagons bonded and interlaced with the 5 indispensable elements, these almost infinite number of interlocking pentagons were in fact what we would call matrices today, and in each other pentagon there would be a special theme, just as original pentagon matrices had the theme of chemical indispensable to life on our planet, so other pentagon/matrices would have other scientific themes, for example the special pentagon theme of the human organs of the body was five essential organs like lung, heart, large intestine, liver or the pentagon matrices for human psychology potential disorders were grief, anger, worry, fear, for culinary taste sweet, sour, hot, spicy, salty.

Each of these five posts [elements] in one sphere would be held up next to an exact analogous post in another sphere. Meanings would be adumbrated between spheres sharpening and made more precise in given contexts. In this way elements in one scientific sphere could be continually calibrated and adumbrated according to context and served as a valuable guide to practitioners in every field, from weather forecasting, to surgical intervention, to psychotherapy to matchmaking.

While this form of practicing applied science might seem specious and unethical to some, its practicality is undeniable. Already its practice is wide spread, not merely in the soft sciences of civil litigation but also in fields as precise and specific as finding the right antibody to antigen model in search for cancer cures.

Regressive physics

CHAPTER 3

While there is little material on theoretical physics in 'egghead' physic of the west outside of some very basic tenants such as plus an minus valences of electrical charges of elements and minute descriptions of properties minuscule particles that compose matter, atoms and their further sub atomic particle divisions and all their properties such as mass, energy, spin and further minuscule properties of these minuscule sub atomic particles such as a variety of quantum numbers. However there are no etiologies or ontologies worth speaking of beyond the assertion that there are four forces, the strong, weak, gravitational and gravity that govern relationships between all masses and exergue of the universe. That these rules consistently govern these elements always in the same way leads our modern western physicists to proclaim the 'standard model', but there is no dynamic theory linking one set of forces to any other, no etiology, ontology so that we are left in a rudderless theoretical vacuum, making the incessant particularization of more and more minute particles and their subdivisions happen or necessary.

As a kind of a backdrop to this achingly vacuous plan of the universe modern physicists give us a defector theory of the origin of the universe called the big bang' which proposes that from an enigmatic welter of nothingness a heavy concentrate of energy/matter appeared, as suddenly struck with enormous external force, split and diffused all manner of this mysteriously appearing matter energy into all forms of matter and energy appearing in the universe today.

This exotic theory, with nothing more than the book of genesis in the old testament to corroborate it, was meant to explain the phenomena of 'background radiation'. Plethora of fields of neutrino that can be dated till supposed origin of universe 14 billion years ago, with decreasing intensity since thus special moments or million years ago.

Likewise in astrophysics of the west we are left with qualitative descriptions and numerical data in detail where it can be found by our more and more sophisticated telescopes data collecting satellites and planet rovers or intruders.

Yet modern western physicists still accept with unquestioning obedience newton's 17th century law of gravity f1=f21/d2, that any small we celestial object is irresistibly attracted to any other celestial object of greater mass with force equivalent to done over distance squared between them. Even though this archaic formula precisely mirrors the degree of subservience of peasant/serf to lord/owner in the feudal systems of the European Middle Ages or the rightfully believed subservience of man to god in the ancient world.

As a result of these antiquated gravitational law of forces in the macro world we are left with data that describes massive supernovas, constellation, global clusters and suns on the one hand, all of which radiate and extrude vast amounts of heated light and black holes, neutron stars and brown dwarfs on the other hand into which the heat and light of radiating bodies are poured--and disappear--all with a sense of meaningless vacuity although recently there has been attempt to diffcrentiate these two types of massive astronomical bodies into 'light mass/energy/ on the one hand and 'dark mass/energy' on the other, there has been no attempt to divine their relationship to each other or to any other physical phenomena in the universe.

When we make a brief review of the working of the 'four forces' however we will see recent findings in physics and math produce so many contradictions in these forces and their workings as to call into question the very basic premises of the 'standard model' which enshrines them as gospel. We will then take the leap from modern western advanced physics to the less renowned but probably more workable and realistic 'regressive physics' of ancient Chinese science.

For example much touted ;strong force' hold that particles ad sub atomic particles of the atomic nuclei are able to defy the same charge repulsion force of the electric magnetic force and dwell together peacefully in the same nucleoid shell of the atom. Thus as many protons as called for by the atomic number of the element can dwell together in the sane shell. This is in spite of Paul's exclusion law which cautions that only one proton plus neutron can live in the sane nucleus, obviating the possibility of a stable atom living in the same nucleus.

Also discovery of radioactivity by polish French researcher Madame Marie curie has sent dilemmas through the physics community still far from resolved almost a hundred tears later

Madame identified the phenomena of radiation as part and parcel of the inherent instability of the atom. While the proton/neutron coupling in the nucleus of every atom seemed to model a kind of sexual mating pattern experimentation and experience has shown over the last 100 years that this atomic coupling was anything but a stable relationship.

Constantly in a state of incompatibility the proto and neutron would constantly wrench apart. Yes the disassociation, separation or divorce that occurred would release important quanta of free energy that could be used, say, in the center of the sun to crate the radiation and heat to light up and warm the universe. But when it happened in a more limited space, for example in the biomass of the human body, it would also be able to disrupt the coupling in adjacent atoms with result that undirected free energies of nuclear break up would cause growth of deadly cancer tumors.

While advanced physicists continue to assert that in the all-consuming heat of the near petal energy making machine of the sun's core, it was never the proton that delayed to produce the generating free energy but the particle that was his female companion, the sexually neutral neuron with a heavier mass than his own.

However the ingenious discovery of sub atomic particles quarks by Dell Mann and Zweig in 1964 would call into question this sexist self-serving stereotype all the fundamentals by which modern western advanced physicists had misinterpreted laws of the physical universe. As the 6 quarks had no sexual determinant nor sexual terminology and were referred to as u, up and down, t, top, bottom, b, s strange, c, charm, and the quark arrangement of protons were uud, of neutrons asddu. The + and - values of the quark no longer referred to sexual like entities but purely numerical worth. Sexuality itself had become a kind of energy arrangement, protons could change to neutrons or vice versa according to energy demands and needs. Sexuality as such as seen through the eyes of egghead elitist sexist advanced physicists was no longer an issue in modern western nuclear science in the west of the late 20th and early 21st venture and like it had always been in the atomic science of the ancient Chinese, was a game of pure power arrangement, not sexual privilege, a game of go, not canasta and while there has been little progress in developing theories of advanced physics in astronomy, western obsession with further and further particularization of minuscule sub atomic particles has left some results in a theory of nuclear energy and mass. Radioactivity, discovered by Madame Curie in early 20th century, has led to many postulates and relatively undeveloped hypotheses about forms of malignant, toxic matter called 'radioactive matter'. While some forms of radioactive matter can be found in naturally occurring elements such as radium, also discovered but Madame Curie, most occurrences of radioactive matter found and studied since beginning of 20th century has been induced radioactivity, induced either by naturally occurring radioactivity in ordinary matter or induced but that ordinary mayyer turned radioactive and inducing the same in new ordinary matter. This study of chain reacting induced radioactivity has also, at the instigation of Curie, been studied for its noxious effects on living things, esp. the human body where chronic plague of cancer in humans was first investigated but curie as the result of induced radioactivity in human biomass. Madame Curie also pioneered use of radiation medically induced to tunic ate random radioactivity within human biomass or at least halt its deadly path.

Most of public knowledge bout radioactivity comes today from widely publicized facts about the massive damage to humans through radioactive weapons like nuclear bombs used against Japanese civilians to end WWII for the allies. More knowledge has been disseminated by use of nuclear reactors used by governments to generate massive energy loads to civilian populations for their energy need it is a tenet widely held by governments and people of both developed and developing countries that use of nuclear energy for people's energy needs will spare environmental and ecological damage caused by burning carbon fuels for these purposes, though the effects of radiation poisoning escaping into the atmospheres, seas, and lands proliferating radioactive particles resulting in increased cancers in all living things have yet to be assessed.

CHAPTER 4
EFFICIENCY OF YIN YANG DYNAMICS ON BOTH COSMOLOGY AND DNA

© Copyright by Eleanor Morris Wu, Alexandria, VA, USA, 2016.

While modern western physics has a scarcity of ontology, teleology in its explanation of the physical world and all its structures, it would be well for it to pick up some of the surfeit of these much needed aids in understanding the physical and natural world from ancient Chinese science and the Chinese Ching. While the terms of ancient Chinese science yin [female] and yang [male] have been adopt into our western vocabulary for hundreds of years, actually since our first contact with this rare civilization from the time of Marco polo in the 12th century, it is only with the findings appearing in modern western civilization in its flowering in the age of experimentation after the age of the Italian renaissance that these terms reached to us their vast scientific profundity and usefulness.

Only with our relatively recent understanding of suns, super nova's, global clusters and constellations and the mechanics within them that produce the heat an radiation to light and warm the universe does the tern; yang; which in Chinese in its rightful context means masculine radiating power, all expanding power to heat and to

Light all in its vicinity to inseminate and cause to flower' finally find objects in the physical world suitable to its meaning. Similarly the tern 'yin' which means, its context, 'feminine ability to conceive and conserve the seed of the flower, to harbor and preserve it, to make it grow but never excessively, to clean up all the dross it spreads about itself in order for other seeds to grow peacefully and unharmed' does the term find its proper object of referral in the physical world as black hole, neutron star often thought of as the vacuum cleaners of the universe as it does away with all excess and waste, allowing space and time for new 'yang' objects to grow and flourish in.

Turin and yang as cosmological forces have their own built in teleology and ontology, they are as it were sexual drives and forces of the physical world. As in sexual forces of the natural world these forces are compellingly attracted to each other while at rage same time, mutually exclusive and even under some circumstances mutually repellant.

While it is true that because when super nova's, sun, constellations meet up with black holes, neutron stars in deep space they come to horrendous conflict in which one of the other is consumed or destroyed, westerners will maintain the mutual exclusivity of these two forces. Chinese however will maintain their complementarity.

On a at once a more limited scalp but also at the same time a wider scale, yin and yang are forces that govern all elements of the a prior neither yin or yang they may have secondary yin or yang characteristics, for example all even numbers and chemical elements with plus valences have yang as secondary characteristic while odd cardinal numbers or chemical elements with negative valences have yin as secondary characteristic.

As we know from work done on DNA by Watson and crick and their followers there are the four de oxy nucleic acids found in DNA that line up on strings that fold into a double helix shape. These four nucleic acids are guanine, adenine, cytosine, two which are purines and two of which are pyrimidine. As DNA is double stranded, and the left sided strand is the DNA in the genome which actually produces the genome whole the right side of the double helix strand is the RNA o template of the DNA which is used over and over again to replicate DNA and send it into cell nuclei throughout the body.

Across the bridge of DNA and RNA, a purine nuclei acid on DNA will always link with its pyrimidine mate on the RNA strand. Similarly any pyrimidine nucleic acid on the DNA strand will link with its purine mate on the RNA strand

The nuclei acids and their mates are as follows.

Research by Watson Crick and their followers determined that it was the sequence of nucleic acids on the DNA strand in units of 3 which determined what amino acids would be produced but the DNA. It was groupings of amino acids in an array of clusters that made up the vital proteins of the organism's body.

Yet the Watson crick model left us no way to predict the number or contents of the amino acids produced by the nucleic acids in groups if 3 that would make up the vital proteins of the organism's body, using ancient Chinese science and ancient Chinese in Ching we may open up new pathways to gaining this knowledge.

The ancient Chinese in Ching, an array of binary opposites in units of sixes or hexagram for a total of 64 hexagram. The binary units are short straight=t horizontal lines and short straight between horizontal lines which we may understand to represent the binary of positive versus negative valence of chemical elements of atomic elements on the periodic table, id these binaries can be seen to represent the 2 groups of nucleic acids, the purine and the pyramid dines there are 4 nucleic acids altogether as there are 2 different purines and 2 different pyramid dines. Thus 2x2=4 means that 2 to the power of 2 yields a quantity of 4 nucleic acids in unit's o2 or in code units of 2.

If we want get the nucleoid array in code units of 3 we must use 2 to power of 3 to yield a quantity of 8 units where units are code is three units. Finally to get the in Ching array of 64 code units where code has 6 units we get 2 to the power of 6 yielding 64 total code units each with a distinguishable code unit of 6 charges or binaries.

But from our previous argument the subject does not end there. Because all code unites squared at least once what were negative binaries in one unit will become positive in another code unit. What was yin in one code unit will become yang in the squared cod unit Heat was hydrogen, -, -1 in one code unit will become hydrogen +1 in another code unit, what was nitrogen-7 in one code unit becomes nitrogen +7 in another code unit. Similarly phosphorous-9 becomes phosphorous +9 in another cod unit.

Even those element that already had yang or + valences may transform when they are squared. This would be a matter of the realignment or adjustment of the quarks in the elements as they found their own best realpolitik. Thus for example in the oxygen atomic number 8 had it's already yang positive volume squared. What would it become? Adding another+ valence to its already +8 valence might yield the gh+2 to add another atom of h2o or water or it might succumb and v=become the unreal noble gas of 10+/ how all these atoms changed under the stress of increased positive valence would be the real politic adjustment they come to under the stressful circumstances.

At any rate we see that an array of three nucleic acids along the string of DNA said to produce a specific amino acid mat not be that simple after all as a possible 3x4 different arrays of 3 nucleic acids may be produced by any string of 3 nucleic acids lining up on the DNA strand due to yin yang dynamics of DNA.

But to finally resolve the power struggle among the atomic element that will determine structure and function of critical protein or proteins comprised of contending amino acids are large structure function of the critical protein in relation to organ system it is instrumental in building. These larger purposes of proteins within organs are adjudicated by structure function of critical atomic elements used in building the. On the next chapter when we explore number theory of cardinal numbers and atomic dynamics of elements with these cardinal numbers we will see the multidimensional , multifaceted influence of structure and function of cardinal atomic numbers reaching into final structures and functions of the . organism itself

Carbon 6.

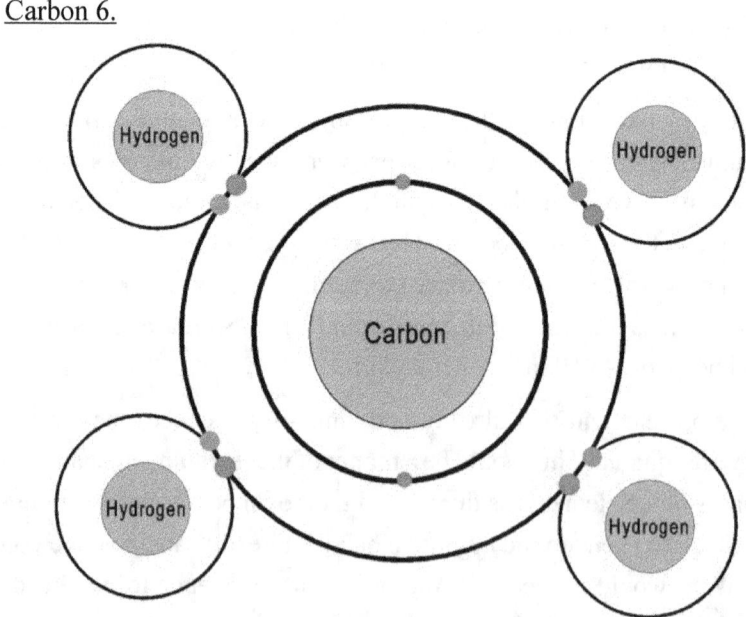

Oxygen.

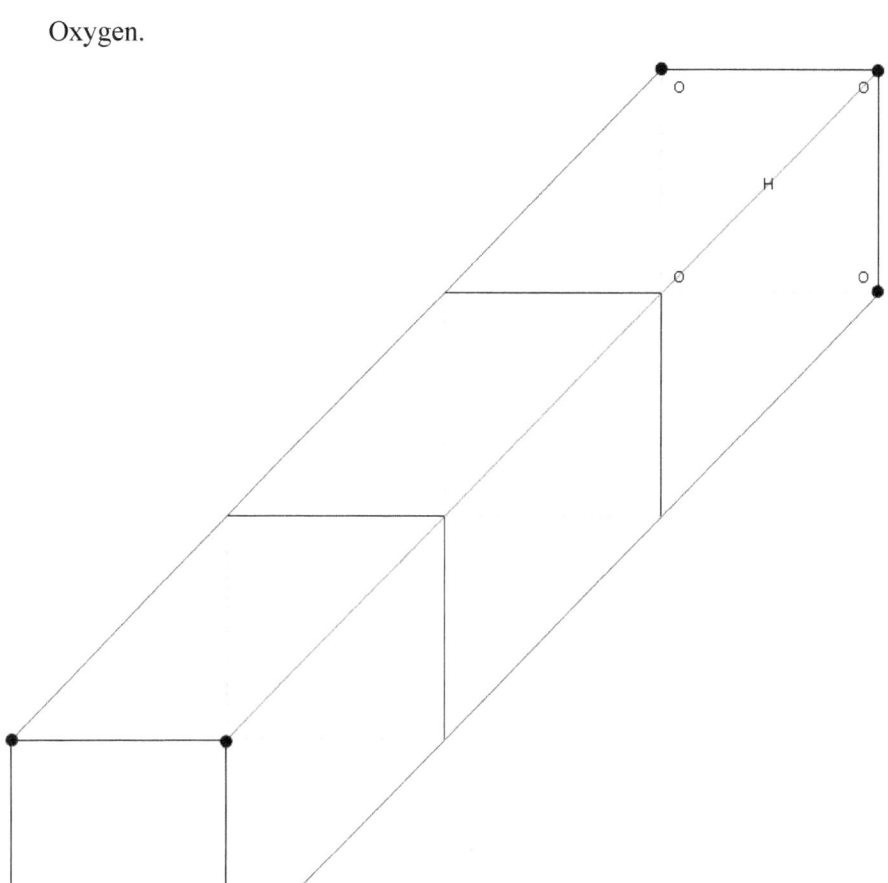

Phosphorus.

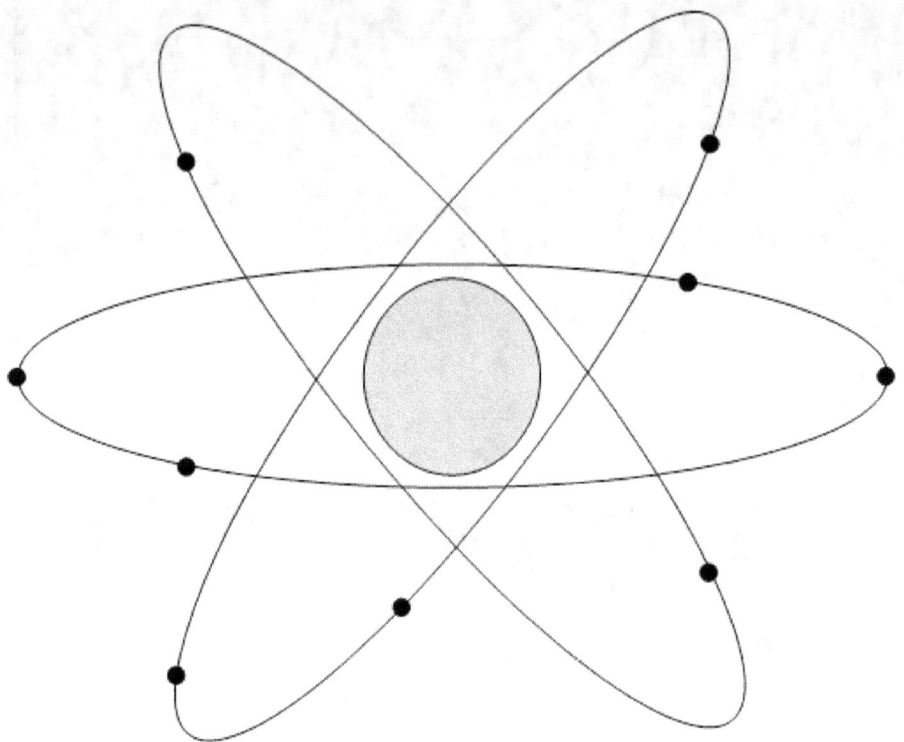

Carbon 6 / Template for Nitrogen 7.

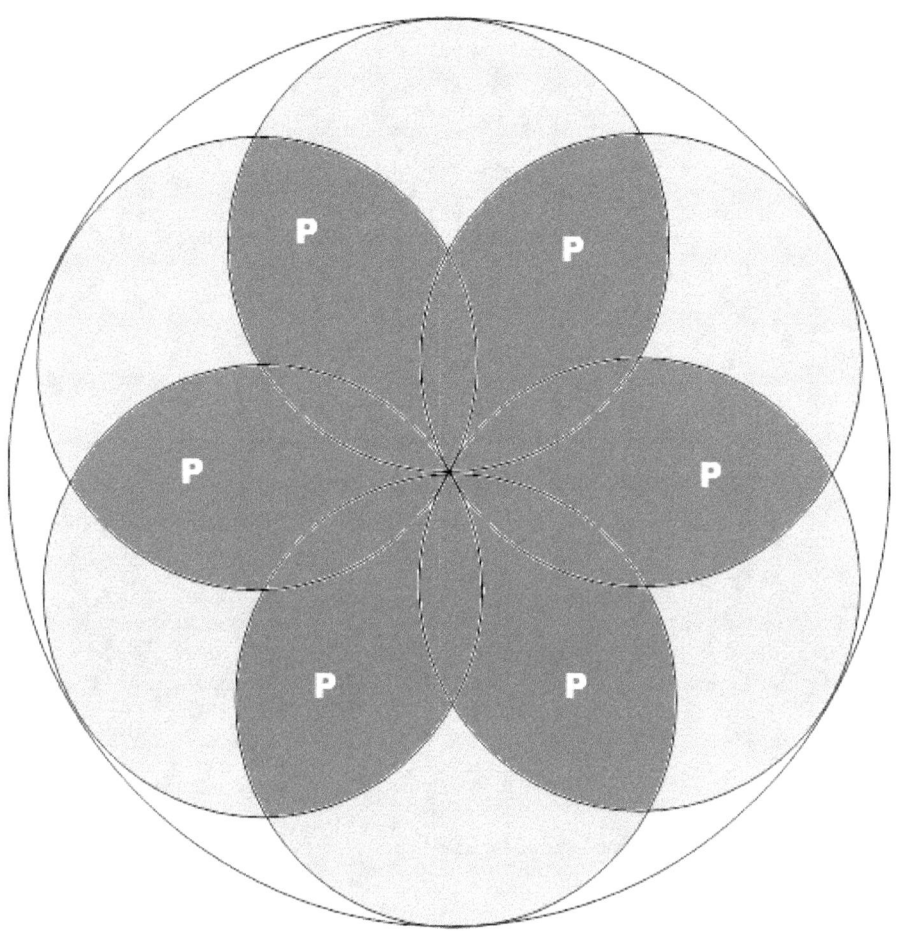

CHAPTER 5
‛ANCIENT CHINESE SCIENCE VALIDATED BY EIN- STEIN'S THEORY OF GENERAL RELATIVITY

© Copyright by Eleanor Morris Wu, Alexandria, VA, USA, 2016.

While ancient western mathematicians and geometers obsessed themselves with the paradoxes of zero, after the Italian renaissance when westerners of the modern era made their first indelible marks on the history of science, their genius cascaded with the near simultaneous discovery of calculus by German Leibnitz and Britain Isaac newton, who were able to envision the circle bisected into increasingly minute right angle triangle whose area could be determined by the Pythagorean theorem and then summed to achieve the total area of the circle. With the invention of engine and other industrial machines in the 18th century, calculus readily found practical application in the industrial revolution of the 19th century. Industrialized weaponry in the 10th century, becoming more and more sophisticated in the 20th century also gave calculus a greater welcoming home. From Ernest Rutherford's discovery of the atomic nucleus and the proton in 1911, atomic physics was born, and the work of Bohr, Heisenberg and Einstein later in the century on nuclear science resulting in the production of nuclear reactors and nuclear weapons gave calculus an even more welcoming home.

By the end of the 19th century with the identification of the atom as the prime identifiable and measurable unit of matter by John Dalton, different chemical elements could be measured and weighed and the science of industrial chemistry was born.in 1869 Russian chemist.

The ultimate clarification between atomic weight and atomic number was not made till Chadwick discovered the neutron in 1932 though considerable work unto that tie had been done on the subject by Dalton, Rutherford route, de biek and Mosley. The first periodic table of the elements was published in 1869 by Russian Dmitri Mendeleev.

When Einstein first conceived of the principle of relativity, he was guided by his teaber and mentor Hermann Lewin Minkowsky. The inventor of 'Minkowsky's space times'. These were no more than patches on a two dimensional graph with three space coordinates and one time coordinate. By making space time an indivisible coordinate Einstein was able to deal with a continuum where space and time could be re-calibrated, even in the presence of relatively high mass and energy vectors. Space time could be imagined to extend out into infinitely large 2-4 dimsionsional Euclidian space. Blips in the continuum could be flattened out and space and time coordinates could be rendered as scalar coordinates even when the gauge difference between them was of a higher power differential. Using this method Einstein was once again able to prove validity of the standard model where all four forces of energy, strong, weak, electromagnetic and gravitational, gave the same results throughout the system as long as they had in the first instance properly differentiated from each other. This was Einstein's theory of special relativity.

However Einstein found when compact objects of bright mass exuded largest quantities of radiation and light or huge dark objects with great density was found in the time space continuum Einstein found that re-calibrating Minkowsky's space times no longer gave the correct solutions. Gauges and scalars could not be adjusted by merely lengthening or shortening coordinates in Euclidian space and Einstein found it necessary to resort to Riemann geometry and basically scrap newton's concept of gravitational 'force' altogether.

In Einstein's 'theory of general relativity' 'gravity' was no longer regarded as a 'force' but as the very landscape of the space time continuum which was being egregiously disturbed and altered by high cyrvitured and extremely massive hot bodies and extremely dense col dark bodies with intense inward curving geometries.

An object of high velocity in this universe no longer travels in a straight line towards on object of greater mass in Euclidian space, rather it travels on a curved line along the geodesic curved space of the pace time continuum towards an object of complementary contours.

While this may seem a strange universe to the denizens of classical western physics, to the wizards of Chinese science and Chinese in Ching bit is just down home stuff

On the universe of ancient Chinese science and rage Chinese in Ching-all mass and energy are subject to the sexual-like energies of yin and yang. Yang the bright spherical masses which radiates light and heats the universe so complicated by the dark yin matter of black holes and neutron stars which protects the domains of space and time by consuming the excesses of yang from the universe, permitting it to thrive and grow with manageable yang [suns, global clusters and constellations] without breaking down the very boundaries of space and time.

While Chinese in Ching is governed by these two forces, known in their familial personae as father and mother these great powers have their manifestations in the universe of space/time, mass/energy which in familial per soma are husband and wife two hexagrams. Further particular stations or manifestation over in familial per soma as older brother/younger brother and older sister/younger sister. While manifestations of husband. Wife on the earthly terrain art fire/water, older brother/younger brother art mountain/stream/

These manifestations of the prime powers of the universe can be seen as first partial derivatives of the prime integers and second partial derivatives of the prime integers.

The setup of prime, secondary and tertiary powers on Chinese in Ching reflects the belief of these ancient sages in a steady state picture of the universe. Yang positive, yin negative, fire positive, water negative, mountain positive, stream negative sets up the cascading of eternally complementary forces in their integer phases and in their derivative phases. So they are set up in continual change in their valence phases from life plus to death minus and rebirth from negative to positive valences. There is complete revolution of all math in integer and derivative, plus and minus phases, a rotation always fulfilling and never ceasing.

Although Einstein favored steady state theory of creation, his instincts were confounded by the existence of ever decreasing background radiation which seemed proof positive of their having been a 'big bang' of quintessential energy mayyer sludge some 14 billion years ago, dispersing matter and energy into the phenomenal map of supernovas and quasar we see before us today.

Yet there is no reason not to believe that 'steady state' and big bang' could not exist contemporaneously if not simultaneously in phase. Luke a pulsar universe could pulsate in steady state until at some point there is no room for expansion and like a giant black hole the inverse suffers gravitational collapse, yielding a combinative sludge of mas and energy remaining in the universe. As steady state forms again, start to pulsate, the steady state of universe begins to reform but now all combinations and permutations of yin yang in the specific particularization of matter and energy, will be different from the previous structure of universe in steady state universe prior to recent big bang.

As predicted by ancient Chinese in Ching and validated by Einstein's general relativity, the universe is a steady state, pulsating, rotating, ever changing in its particulars but not in its general dynamics.

——

CHAPTER 6
SQUARING OF THE CIRCLE

© Copyright by Eleanor Morris Wu, Alexandria, VA, USA, 2016.

In ancient Greece the Paradoxes of Zena in which centuries of Socratians and their opposition Sophists sank deeper and deeper into the kind of linguistic quagmire we still struggle with today in all cultures whose languages are patterned after the Phoenicians. Based on letter symbols memorized according to sound sequences that accrue in sequences to make meaning codes or words whose adjusted sound has an entirely arbitrary relationship with objects that signify, possibility for incomprehensibility in the language is infinite. Because words, made up of sound bites or letters, have no significance in representing the objects they signify besides an arbitrate code given to memorization the language speaker is constantly disassociated from the objects he purports to be speaking about.

Phonetic languages lend themselves easily to madness on both the individual and collective. Compensating these cognitive allays is these of processing digital data between individuals and between individuals and collective, given that Phoenician language are essentially no different from programmed binary codes computers use to converse with other computers.

The kinds of written and spoken languages which stand in opposition to phonetic codes are the digitalized picture languages of the Egyptians and then on digitalized impressionistic picture language of the Chinese. While Egyptian picture language adds information in their script of realistic pictures the written script represents, in the Chinese language figure like impressions of the objects the scripts are set to recognize are presented

While ancient Egyptian language is a semi digitalized, semi phonetic zed language, Chinese represents a language of concepts, presented impressionistically and understood holistically or comprehensively.

Communication between collective to individual will be the most common with individual to collective much rarer since relationships between individuals and collectibles would regimentally solidified by any message from the collective.

As the language is almost entirely devoid of digitalization, communications between collectives and individuals in monetary, trade or technology matters would be difficult although there would be more leeway between individuals. In all in general in this Chinese society knowledge of a secondary digitalized phonetic language or robust use of binary mathematical [in recent times, electronic] language would need to be robust to fulfill major needed economic transactions, technological or military enterprises.

Among the ancient Greeks, with their lust for science and mathematics and their devout belief that these were both branches of logic and philosophy, there was a veritable culture crisis as the Greeks attempted to integrate the great practical advances of other ancient nations into a philosophic rubric they could be comfortable with.

Attempting to assimilate the architectural advances for example of the Egyptians, the Greeks had accepted Euclid's perfectly logic geometry as the hallmark of their civilization taking us down to the modern area under its banner of scientific rationalism. With precursor of full algebra later to be integrated into mays and sciences in the West from the Arabs all of Euclidian geometry validated its logical correctness by commutability of lines into squares and squares into solids both geometrically and algebraically.

Thus two straight lines at right angles when doubled or squared would yield a square. Every square could be bisected in half, or halved to obtain the two straight lines at right angles again. If the two lines at right angles were of the same length their squaring would yield a perfect square if not the same length the consequent 2 dimensional figure would be a rectangle.

To get a solid square or rectangle, a volume, simply mores right lines at 90 degree angles would be drawn at intersection of lines for square or rectangle and connected and then also squared.

Problems started to come for the ancient Greeks when they came to Egyptian findings of what became known as 'Pythagoras's triangle, the square root, pt. and the circle.

That the diagonal of a triangle with two straight lines at 90 degree angles could be found but squaring the lengths of the two lines, adding them and then taking the square root of their sum was troublesome enough. [a2 +b2=c2], yielding as it did a decimal number, a quantity with no place in their rational scientific lexicon, but even further knowledge that this troublesome diagonal multiplied by the constant pi. Also picked up from the enigmatic Egyptians would give the area of a circle, cause them and their heirs to Western civilization in Europe, no end of consternation.

Beginning with Greek mathematician Hippocrates of Chios [410-470 BCE] the Greeks attempted to square the circle, to find some way to make its area commutable, Hippocrates attempted to divide the circle into four medium sized squares and compute the area but much of the area of the circle was always unaccounted for. The circle could only be commutable of it could be divided into squares and accounted for--other than that the circle was neither rational, scientific nor real and the Greeks would have none of it.

The problem was finally solved--or at least in part-nearly 800 years later when Leibnitz and Newton nearly simultaneously invented calculus. They both took the circle, divided into in Foote number of infinitely tiny [imaginary] right angle triangles, calculated the [imaginary] areas of each triangle and added their [imaginary] sum to literally square the circle and gets its toper [imaginary] area.

With the squaring of the circle and development of the calculus modern western scientific, technological, industrial, informational, western civilization was born. Westerners went on to discover the atom, atomic weight, atomic number, its substructure the atomic nucleus, proton, the electron, the neutron, radioactivity, nuclear energy, atomic fission, fusion, and the nuclear reactor.

But the dilemma that beset the Greeks 2000 years ago had not been solved, if anything it had become worse for from the position of a Greek sophist Socratian skeptic, all of these wonders were built on not something real but something unreal and imaginary, or in the words of the Platonic skeptic, how did the light get into the cave?

Modern mathematicians, scientists and astronomers are, happily taking a more pragmatic approach to all of this and this pragmatic approach may soon be the real deal of the day when modern science learns to assimilate ancient Chinese science and accept with a lot more humility than the rational Greek skeptics were ever able to do, our phenomenally minuscule part in the universe and everything that is in.

Yes it is true, circles, spheres, our earth, and everything on it, and everything we have built on it, and will build on it, is unreal, imaginary and, in the total scheme of things, magnificently irrelevant to the way the universe operates and what is really important in it, 'oh brave new world that hath such creatures in it' says Shakespeare in the Tempest as young Miranda meets a young male for the first time.

We humans are the strange creatures, we are the anomalies,

Chinese ancient science, corroborated by our own findings in physics and astronomy, tells us the universe is a giant battleground between yin and yang, super luminous suns and supernovas [yang] and super dark dense masses [yin] black holes, quasars, neutron stars. These universal combative sexual energies of the universe, which have next to nothing in common with the sexual energies of living things, let alone humans, would consume our mightiest warriors, our most seductive women, in less than a nano second should any one of them come in contact with universal sexual energies of Yom and Yang.

What then is our function in the universe, or are we truly just imaginary as Shakespeare and other great thinkers envisioned us? While this question in the past has always been answered by litterateurs, prophets, philosophers, then opens and saints, our progress in math and science may now be enough to answer these questions with these too.

First we must look objectively of what we actually are in scientific and mathematical terms, the place we live in and its neighborhoods

Minuscule collection of some few cardinal atomic elements and combinations of these elements in compounds, both carbon organic and oxygenated bio mass, we live on a terrestrial planet that is basically rock. We have one luminous sun, three plants inward, none planets for our sun altogether, where not one of the other planets in our solar system are habitable, at least not for oxygenated carbon computed and biomass life we enjoy o earth.

And what about the sexual energies we have in all living things on earth, the energies that drive us to court, mate, procreate, and care for the offspring we produce and the elderly too eat any longer to care for themselves? And in an even more arcane way motivate us to learn, study and then create the truly ethereal works of art, music, poetry, philosophy and science? And to build eternal architecture throughout the planet, ethereal places, to live, eat, sleep, work, mate, play, raise offspring and care for our elderly? what relationship do these fancies have, even less useful to the workings of the universe than our own unuseful paltry sense, living on these unuseful paltry rock or gas planets with little understandable use in the great mechanics and dynamics of the mighty forces of the universe which keeps it running and in good repairs while we useless ineffable life forms with their ever increasing unuseful products such as offspring, art music, poetry and science loll around in it?

To understand that we do after all have a role, and it is a useful role--however minuscule--for the workings of the universe we must first try to understand the nature of matter and energy in the universe and see where we fit in. We will depend not only on our own great scientific and mathematical findings from our modern western scientific technological; civilization but also from the timeless wisdom of ancient Chinese wizard of math and science who are still willing total to us if only we will listen.

CHAPTER 7
GEOMETRIC OUTCOMES OF
CARDINAL ATOMIC NUMBERS

When there is an imbalance of atomic elements in the nucleus as in nitrogen no7, the odd subatomic particle will pulsate in and out of the shell in the form of a neutron or dark matter wave. When the shell is still coded and cardinal number is not commutable the neutron or dark matter part of each atom will temporarily or continuously protrude in the form of neutron or dark matter wave.

Yin Yang dialectics, further divisibility of sub atomic particulars, the wave-particle duality

The great Hermann Lewin 'Minkowsky, the 19th early twentieth century mathematician and theoretical physicist, had long dreamed of, once cardinal atomic numbers had been found, of calculating the motions of these cardinal numbered atoms and predicting their geometric outcomes. 'Minkowsky had no doubt planned to use the same method of verifying the motions of the cardinal membered atoms as Euclid had used to verify geometric truth of basic geometric shapes by positioning then in space and showing how squaring and then cubing an original three straight lined figure at 90 degree angles would yield projections of the original figure which then could be square rooted or cube rooted to yield the original lined figure, i.e. by commutabity. This time not just the size of lines or degrees of angle could be verified but also the speed and trajectory of motion and energy of the element consumed to propel it.

However while 'Minkowsky used in the late 19th, early 20th century, the atomic nucleus with is chief motive force of the protons. The number of which in any atomic element, was not discovered by Ernest Rutherford until 1911, 2 years after Minkowsky died of a surprise appendectomy in his home in Konigsberg, Germany. Because of his identifying the atom nucleus and measuring the weight of the proton, Herbert Mosley, in 1912, a year before his death in the first ww theatre of galpolie, was finally able to identify proper atomic numbers for all the known atomic elements at that time, in the periodic table of the elements, still basically the same as used today.

Although Rutherford identified atomic power per se, it was not till a fuller understanding of nuclear power, in combination with a fuller understanding of radioactivity, fostered but the work of Bohr, Heisenberg, Einstein, Curie, Rutherford, Oppenheimer, Cranshaw, Zweig, Gell-Mann and others, was a truer understanding of atomic--or nuclear power gained.

While the Manhattan project of the early '40's under direct command of president Roosevelt, marshalled the minds of the best allied physicists to produce a nuclear device to ensure victory in WWII and before the Germans did, greatly enhanced the bread and butter facts of nuclear energy throughout the scientific community, it was not till the work on quarks but Zweig and Gell-Mann in 1964, that the last keys to the mysteries of nuclear energy were unlocked.

Brought to proximity of our galaxy some 14 billion years ago, a portion of a large supernova of hydrogen gas released in the paroxysm of energy and mass distributing due to the 'big bang'. Our sun, has been assessed to have another 4 billion years of life till it begins in enormous burn out, becomes a red giant, and the, after, become a black hole, then dead crisp of wasted mass and energy and disappears into the reaches of stellar space.

In its present, recent past and foreseeable future our son resembles nothing so much as a perpetual energy producing a cjine of the light and energy necessary to warm and radiate our nine planet more solar system and make its presence known in this quadrant of our light year galaxy known to us as the 'milky way'.

Inside the core of the sun among the trillion hydrogen atoms bequeathed from its ancestral super nova, the protons n b=neutrons of these hydrogen atoms are in a virtual process of continual death almost instantaneously re-charged to life again by other neighboring sub particles--other protons and neutrons. In this constant contest of martyrdom and resurrection, the neutron on the back of every proton is stripped clean, yielding her divorce from her proton a vital segment of radioactive energy which when accumulated to the trillionth power of number of hydrogen atoms in the sun core fields the heat and light that powers the sun and heats and lights up her solar system. Meanwhile stripped from her proton mate, the neutron now still emitting the radioactive energy bequeathed to her in her divorce, assaults a neighboring hydrogen atom, stripping the lidless neutron from her proton support in the same way she had been divested from her conjugal embrace with her proton nano seconds previously.

While mass of neutron is 1 and mass of proton is only 99.86% of neutron, the game of mass/energy accretion in the sun's core is difficult to lose, which the sun's mass energy can last so many billions of years without a glitch, for a total of 8 billion years, 4 in the recent past and four in the foreseeable future it is a win win game, making the sun a virtual perpetual energy releasing machine--ubril daddy runs out of cash and there are no more protons to burn ti usable radioactive energy--and finally in the death throes of the resulting black hole produced by a dying sun there will not even be any more neutrons to cannibalize for free radioactive energy. Too light up needy dependent planets.

While the actual constituents of the atomic nucleus, even in the heyday of nuclear research in the Manhattan project in the early '40's was shrouded in mystery, since the work on quarks but Gell-Mann and Zweig in the '60's, most of this mystery has been dissolved.

While since the time of Rutherford in 1911 till the '60's the basic constituents of the atomic nucleus. Proton, neutron, electron, were believed to be inviolate and incapable of further subdivision or change of any kind. The work on quarks done by Gell-Mann and Zweig changed all that as atomic particles of the nucleus, proton, neutron, called hadrons, barons, were found to be themselves divisible into for the proton divested of his neutron companion, the assaulting radioactive 'other woman' , radioactive can send him the boson to convert his quark load from up [proton] to ddu [neutron] producing another particle with heavier mass 1, waiting to be assaulted but yet another rogue neutron and become an electromagnetic plus proton again. The point is in the sun's core, atomic number per se, so important in the solar system where compounds are formed from elements of distinct atomic numbers, in the sun's core atomic number per seis irrelevant--all that really matters is the quark power value of the particles, enabling a game of realpolitik to be played in every contact between the sub particles of neutrons and protons, in this inferno of power politics been sexual identity or yin/yang goes by the boards and is scarified in the interest of obtaining maximum power, i.e. radiation and light--for each contact event between protons with up quarks and neutron seith ddu quarks, all vestige for sexual dignity of individual sex power is sacrificed for total power gain, now it is the total power gain of all sub[particles in the sun's core which yield the massive all yang value of the sun. This exchange of the charade of sexual male power for real power given to heavier mass neutron is made possible by nature of solely power driven quarks which unceremonious-ly sacrifice the previous power and prestige pf yang proton for now liberated naked proton's ability to receive extra mass and become another heavier mass neutron. Thus once the neutron is stripped in the first place the proton she then becomes happily sacrifices hr male proton yang stays for the heavier mass that will make him a neutron, aiding the yang status of the macroscopic yang sun to main his massive power and to carry out his macroscopic yang destiny in the universe. The charade of male power of.

Individual yang proton is sacrificed in the game of real politic which gives total mass yang sun the ability to fulfill its destiny in the universe it is total massive power accretion which produces the sexual yang energy of massive compact bodies such as suns, supernovas, global clusters and constellations. Could be broken down into further smaller subatomic particles with their own array of quantum numbers, called quarks--who were also baryons, hadrons and fermions.

Altogether there were. Like there were for DNA, a six figure array of possible quakes whose number 6, for line of code, was exponent of a base 2. Highly analogous the cod, its power, base and exponent to the DNA code used to determine amino acids and their sequences in the protein that makes up the flesh of all living things.

However in the case of quarks, the six and 12 figure codes. With a base of 2m exponent six, does not determine sequence of amino acids determining all flesh, but sequences of matter. Energy codons making up the physical. Chemical and mathematical flesh of all physical, chemical matter in the universe. In other words the determinant of the 6-12 codons of physical chemical matter like the physical-chemical skein of the universe, lying as a substratum to the protein of all living flesh in all living mater in the universe while the base for the six layered codon of the genetic code , made up the four nucleic acids forming amino acid formations strung together to make protein, so the four bases of quarks are the four fundamental forces of nature, electromagnetic, weak, string, and gravity, making up the 'flesh' as it were of physical chemical structure throughout the universe.

Thus 2 purines, 2 pyrimidine, differential to four nucleic acids, adenine, guanine, cytosine and thymine, connected linearly along the DNA string, in groups of 3's make up the 20 known amino acids necessity for living matter, so it might be conjectured in 'living matter; of physical chemical life the combination of four forces in groups of three make up 'amino acids of physical chemical skein of life, physical chemical skein of life that make up our physical chemical life and is the underpinnings of our biological life.

While it present we do not have the knowledge of physical-chemical basic structures, up pinpoint and analyze them with any certainty, they remain a useful paradigm for future scientists to analyze more closely and call our attention to the truth the ever increasingly likelihood of these paradigms as likely complementarities as laid down by wizards of ancient Chinese science based on heir 5 indispensable elements, fire, water, earth, metal and wind. On my book ancient Chinese science and the Chinese in Ching bring to the attention my readers the indisputable similarity between ancient Chinese indisputable elements and our 5 indisputable biochemical elements of modern times, the pentagon ring of the DNA molecule and the 5 elements which connect it, i.e. ,fire=hydrogen, water=oxygen, earth=nitrogen, metal=phosphorous, wood-carbon.

More exact research into atomic structure coinciding with the work of Gell-Mann and Zweig, has shown us that far from being non divisible entities, the neutron is a combination of the proton plus electron.

Thus in crocus of nuclear decay and regeneration that goes on in the sun, a loose neutron can dislodge an obedient neutron spouse from the back of her conjugate proton husband, the free energy that has dislodged wife neutron from back of husband proton is extremely power radioactive energy. The parcel of radioactive energy used to dislodge nucleon conjugates comes from energy released when rogue neutron was released from her own conjugate, this packet of energy she carries with her as she dies the energy that allows her to release the electron from the faithful conjugate, all these dislodgements release further packets of radioactive energy now carried but newly divorced conjugate and electron on their orgies of dissonance against further loyal conjugal end electrons in the sun's arena.

Yes the once loyal to her proton male conjugate, the now offending neutron can be further cannibalized by wandering radioactive particles haunting the halls of salacious sun, and she can further be riven apart to a singular rote in as her electron which once made her intact neutron is further riven from her.

Thus this game of sexual slash and burn is a win win situation for almost perpetual energy, light and radiation gain--all radioactive for the trillions of hydrogen or even other atomic numbered elements in the sun's core. Through micro machinations the total yang nature of the sun is maintained throughout many billions of years eventually the heavier mass and increasingly radioactive neutron particle prove triumph and and a yin black hole is legacy pf yang sun.

In time numbers of protons giving strength and direction to hydrogen atoms decrease and closers of very massive highly radioactive neutron, evolving in the twisted convoluted way of interior spatial convolution as appears in Riemannian geometry takes over the spatial parameters where the sun once soon. Blackbody radiation is a feature of the black hole as is increasing convolution of recognizable space. Finally all recognizable space seems to disappear in a Schwarzschild radius and vent horizon. Hawking has declared a kind of final thrust of matter as it oozes out of the hole, pops over the event horizon and disappears into space, its final destination if any, unknown.

The combined phenomena of the imposing sun or yang combined with aerially threatening spectra of a hole in space where all recognizable scales makes up the landscape described by both Einstein's theory of general relativity and the steady state repeating pulsars of ancient Chinese in Ching. For Einstein the best line word mem for this universe was gravity' for the ancient Chinese wizards it was 'change'

In trying to solve even more fundamental problems of physics, looking at the universe through the prism of ancient Chinese science and in Ching we see we might finally have a rational solution to particle wave duality that has plagued western physics since its inception in the 17th century,' From the most succinct point of view we may see the appearance of matter as waves as the yin perspective, matter as particles as yang perspective. Tied to cosmological history the accent of yang in the universe happened in the for front of the 'big bang' and reflects rge universe's origins withe appearance and proliferation of atoms with lowest atomic numbers like hydrogen and helium, while we may see advent of Yin in cosmological history as end of days, dying of the race and the appearance and advent of all elements with two or three digit atomic numbers, at the end of periodic table of the elements chart/

This end of day's scario for atomic elements with the largest atomic numbers plays well into our scheme of yin yang parameters which we have discussed partially in previous chapters.

But the fuller significance of yin yang parameters for all atomic elements will be discussed in more detail in coming chapters fryin yang parameters are not just in reference to beginning. End of days but to real/unreal characteristics of all atomic elements

We have already mentioned that in solar system, our very existence as oxygenated creatures of bio mass s unreal in a very general way as not thing we do or no circumstances in which we find ourselves are in any way pertinent to the actual workings of the universe.

Nonetheless within the circumstances in which humans find themselves there will be parameters of more or less significance to the way humans operate as social differentials of truly real or useful prime dynamics in the outer universe. This will take us to serious consideration of entropy as the litmus test of our usefulness for a wide range of parameters for human beings in the solar system.

CHAPTER 8
COMMUTABILITY OF CARDINAL NUMBERED ATOMIC ELEMENTS OF THE DNA MOLECULE

© Copyright by Eleanor Morris Wu, Alexandria, VA, USA, 2016.

Hydrogen atom of the DNA molecule [fire of the five indispensable elements of the ancient Chinese science] has no commutable deficits as it a fermion, hadron, barton awash in a sea of quakes, all plating themselves out the real politic game of power maximization. In the cauldron of the sun's hot core where this real politic is being played out hydrogen is stripped to his zithers, both his proton, neutron and electron components are all dispensable where net pore gain is the only card that matters. While hydrogen before its sacrificial fissures is composed of male yang proton topped any electron lepton, all the while resting on his back in filial embrace a neutron, herself when stripped, an immodest ladies composed of proton and electron. So it can be seen that the entire atom in its least nudist aspects is really two neutron in various levels of dishabituation, in one frenzied environment.

The quark composition of the denuded proton, composed of 2 up quarks and 1 down quark, uud, can be and is shifted any and every time as the laws of chance dictates to the less gender prestigious, but more massive neutron with two down quarks and one up quark, ddu.

The incredible power agility of this ever sexual identity changing power behemoth, will be a chief and highly cherished additive in all chemical element, such for example, as water, as leave the old universe of suns, supernovas and black holes of the massive old universe and travel into the large, but not massive world of the solar system and into even more minute universal phenomena which live off, or on, the planets if these solar systems i.e. igneous rocks, carbon chemistry, the five phyla and man himself [with all the ever more minute life forms who feed, thrive and grow on major-minor life forms on the planets, such as bacteria and cells.] Hydrogen as a combined additive has extraordinary roll, along with other elements of the DNA molecule such as carbon, oxygen and nitrogen as being one of the great additive elements so instrumental to the life form which populate planet earth, so fruitful it has, long ago been memorialized by her human inhabitants as a lush garden, the garden of Adam and Eve, often thought of and duly recorded as the first human beings, male and female, ever to live on earth and denizens of its garden.

While as a single atomic element hydrogen with atomic number 1, hydrogen lives a charmed life, being designated ti join all elements of the DNA molecule in a union of complementarianism, long before the abundance of living phenomena it helped bond and grow, were comprehended. While hydrogen in her role of atom, broke every rule of serial primacy, in his role as element, the escapades he lolled in as a simple atom was never enjoyed by him when he was an element, as a fully functioning much in demand element, hydrogen was, for the most part honor bound to maintain the dignity of his male yang sexual gender and he was duly controlled an regulated by a plethora of strict rules pertaining to that gender; he was no longer hits a role of the dice as semblance of quarks that could be changed every moment as real politic demanded, in his role as element with cardinal atomic number 1, he was mandated to sticks to a strict male additive identity, part and parcel of more down chain derivative identifying the world of mathematical means parcel of life forms and organic compounds fund on the planet since early days when all matter and energy formed in faceless smudge, bound together to create the dispersion of matter abdenergt in one of the pulsations of the semi state balance of the universe.

In its dimension of the universe these strict rules held and the previous days when hydrogen could revel in his quark bath never knowing what form of atomic element he would turn into next, or for how long, as a compound additive hydrogen the element was bound to maintain his rightful form on pain of painful death an organic decay.

Even though two hydrogen often acted together as additive yang, 2 hydrogens lying together in apparent conjugal like mutual solace, nothing could be further from the truth, as the conjugation of two hydrogen atoms together non commutable and were highly unreal to the point of bringing unreality itself as a characteristic of the element itself and further to any element of a third or fourth dimensional element product as in the compound of water composed of commutable oxygen and non-commutable.

Two hydrogen atoms lying together in the same shell as h2 are deemed impossible not Pauli's exclusion principle which deems no 2 fermions with the same quantum numbers can share the same atomic shell simultaneously, thus the h2 of water h2o I deemed unstable giving water is very feel real sense of unreality. While hydrogen with atomic number 1 is the most abundant element in the universe with 71% of all matter in the universe, no 2 most abundant element is he2 with 24% of all matter in the universe.

Yet like h2, he2 is an extremely volatile and unstable gas; it is the second most abundant element in the sun's core and like hydrogen itself is an easy customer of the quarks who tear it down and build it up at will, all in the service of the power politics of the real politic that is the sun's name of the game.

Yet as well as being the second most abundant gas in the universe, the second most abundant element in the sun's core its instability, the ace with which it can change its identity under the pressure of real politic is highlighted by its atom number is 2, which makes it a non-commutable element, a roving meme, as it were in the sea of half whole elements, gases, metals, acids which make up the ocean of atomic elements in the universe, randomly dispersed on cosmic paroxysm 14 billion years ago in the big bang. Indeed outside of hydrogen and helium all other atomic elements combined make up only 2% of the universe. While 2 as a number and a 2 dimensional geometric figure may be commutable in Euclidian space in two dimensions, fir an atomic element which exits in 3-4 dimensions it is not.

Like other non-commutable elements with cardiac numbers, 3, 4, 5 atomic elements, they will appear as half atoms', ghosts or memes in space, thus beryllium, a rare atom with atom number 4 is also non commutable as 3-4 dimensional atomic element

Lithium, with atomic number 3, the magic number of unreality, a highly reactive and intangible gas, not surprisingly has found efficacy in the medical field as a drug controlling symptoms of schizophrenia and depression, all mental disorders where the sufferer has difficulty separating the real from the unreal, realty from fantasy.

Boron with atomic number 5, one of the magic collector' numbers along with 2, 3, and 4, is that magic; collector; number of the 5 sided, never closed circle, the pentagon whose shape is so instrumental in all biochemical science and DNA, RNA, the pentagon as the primary shape of nucleic acids, eukaryotic cell structure, theta cycle of the mitochondria, many amino acids, proteins and appear in critical hormone and neurotransmitter structure. Appears to be, in oxygenated biomass biology, te ame importance the screw has upon mechanical engines, all purpose, all useful, indispensable. Also not surprisingly the ancient Chinese regarded this five loop wonder, the repository of the complementarity that mjes the universe work.

Oro, atomic number, boron, is only found in outer space but forms important compounds like the other non-commutable atomic elements with cardinal numbers, 2, 3, and 4 and is important in making calcium ingestible by bone and in preventing osteoporosis.

While lithium atom no 3 is double 6 [2x3= is, so the number three chief determinant for unreality in the universe, an so central to every kind of life on our planet, in its doubled form as carbon, atomic number 6, sets the wheels moving for life in all of its manifestations for human life in our planet, in our solar system and in the universe, to come full fruition

Caron, wood in the Chinese indispensable elements is the base of organic chemistry in our world, indispensable in producing the foodstuffs web eat, to the majority of the amounts that make up our earthly form, from bone to blood not only that the round circles that are the constituents of the element carbon represent the shapes that most of our bodily features are programed to take from our DNA: our circular body cells, the shape of our circular elliptical skulls, and al skein of our body organs from circulate heart, to heavier circuit bore, stomach, intestine. When we look out the window day or not most shapes we see are predetermined by the number 3 or its cohort pi, from the stars we see in the sky not night moon there positive valence, whatever way we look at the element nitrogen it is always in proton neutron part too much, gigu=giving a b=negative valence or one proton neutron pair too little giving it's still a negative valence. If we may look at our graph of the atomic elements and see them as circles, we will see nitrogen with its extra proton neutron pair is jutting in and out of the circle enclosure attempting to get some closure on the ill-fitting element landscape.

In fact the main determinant to; living things is their carbon skeleton, whether it be the lowliest bacteria, the rarest flower or plant, to the giant dinosaur.

The next three element with their low atomic number/cardinal numbers pretty well fill in the details for lie on our plane as we know it. Valence or as a full carbon atom of 8 with an extra proton neutron added on to give it a Nitrogen with its atomic/cardinal number7 is also deemed commutable as a carbon commutable element with its extra proton/neutron giving it its negative valence. Similarly as an imperfect oxygen atom minus an extra proton. Neutron pair it may be receiving this extra needed proton. Neutron pair from a nearby afflicted atom, receiving it and emitting it in repetive thyrsts and par as a pulsar. The main images we get from nitrogen in our mist is one of movement, thrust and par, along a straight line, giving us in our DNA program the instructions for all mobility movements as well as for pulsar like mob=elements among bodily organs such as heart beat, bowel movements, insemination thrust, labor pons, delivery contractions.

Carbon 6 makes lithium 3 and all similar unreal commutable by doubling the 3's in 2nd dimension and then extending them into 5rd-4th dimension, dibbling of 5 takers atomic no 3 in 2nd dimension but intake number 3 then doubled takes the element into 3rd dimsio due to the fact that initaielement-was3 to start with.

Oxygen 8 with a final positive valence is the ultimate justification of atom numbers 2m4, finally cubed and yielding a 3rd dimensional atomic element commutable on Euclidian space.

Oxygen 8 is essentially important for mammalian and human life because the commutable oxygen 8 provides a time space position for the oxygenated biogas to be placed in a clear real position in Minkowsky space time, oxygen 8 as providing Euclidian dimensions to the life form cannot be underestimated. While carbon 6 provides the cells and organ bags or containers for the life forms many differentiated functions oxygen provides the viable space timrectent the life form possesses and carries with him as his essential life guarantee. This extension dimension is underscored by breadth of lung space and broad expanse of various body parts through which continual broad swathes of oxygen are carried by oxygen infused red blood cells that course in perpetuity to the oxygenated viol masses' lifespan.

After o8 comes phosphorus 9, the metal of the 5 indispensable elements of ancient Chinese science and the sole metal component of DNA, the clock-like precision three minor circle of atoms in threes alternate protruding from the individual cell of 3x3 smaller circle shells that make up atom content of phosphorus 9, this ability to show and hide with clocklike precision mimics the oxygenated bio mass's endemic harmony with sleep patterns of all life beings in the carbon circle, it demonstrates his ability to miicvarious stats of awareness and slumber within the cycle of the larger cycles of all carbon skeleton beings within the larger life cycles of rural mass harninies and conflict behavior common to phenomena preceding and outside of regional suns and their derivative rocky planets or gas, ice giants.

——

CHAPTER 9
THE BLACK SUN

© Copyright by Eleanor Morris Wu, Alexandria, VA, USA, 2016.

In order to understand the major forces at work in the universe, it is necessary to examine these forces from both microscopic and macroscopic points of view, Looking at the machinations of particles in the makeup of supernova global clusters, constellations, suns, powerful yang forces with to that in the central circle of the sun there is battle of microscopic proportions going on between subatomic particles of atoms, namely the proton, the neutron and electron, As the proton's electric charge is plus, the neutron's is 0 and the electron's s minus, we quite quaintly give these particle sexual identities akin to those in animal, esp. human life on the planet earth in our solar system. As proton's biggest opposition comes from neutrally charged neutron rather than negatively charged electron, we assign the neutron also a feminine identity along with electron, perhaps feminine in the mother role while electron is feminine in the wifely role. Yet we know now that the neutron is a composite particle, composed of proton and electron.

Discovery of quarks but Gell-Mann and Zweig in 1964 unmasked certain other features if sub atomic particles, the sub atomic particles in the sun's core all had one important mission--grow on mass in order to grow the total mass of the macroscopic compact body, the sun. How this was to be accomplished was to unleash the radioactive energy from composite neutron particle. This released radioactive energy allowed more attachments of the less massive subatomic particles, the proton and electron to each other producing heavier particles like neutrons, or using the released radioactive energy from neutron breakup to cause breakup of their more simply attached neutrons.

In the game of real politic going on in the sun's cauldron, the quaint typing and matching of heterosexual partners soon proved a sham as power was the central object of the game and quaint heterosexual matching's were no more than fake marriages of convenience. Yet however cynically the matching between opposite sexed sub atomic particles was, their very frenzy in performing their naked liaisons had a further agenda; the accretion of more and more energetic bindings in the service of their macroscopic master, the sun. The more energy the microscopic bits could generate, the greater the power of their macroscopic master the sun. For all intents are purposes the increased and ever increasing power of the sun was used to radiate energy and light: its main object in the ken of our sun and solar system was to warm up and light our solar system. At least from earth's point of view this meant the capacity to generate life in earth looked at from a purely objective scientific point of view the life generated by solar power on earth produced life forms, also in two genders, that could procreate themselves and produce a myriad of produce totally unrelated to the processes that produced the solar system but could be seen as surrogates or differentials that too needed energy to survive yet this level of energy was at a far lesser magnitude than the energy of the sun to produce its first offspring of oxygenated biomass life forms.

In other words the purpose of generating a solar system that could produce life forms was to enable these life forms to produce surrogate life forms using far less energy than it t took to produce them. This general downsizing energy expenditure while still produce products as surrogates to life forms could be the summa of universal motives--to produce surrogate life forms while still conserving general energy of the universe. Since entropy is conserved energy, it can be thought the ultimate pulse of the universe is to protect and conserve entropy.

Thus in a real life scale in millions of poor farmers or hunters run around on earth to obtain the provisions to feed their dependents, they will accrue enough profit to pay taxes to the kid--or present govt. This present government may have a sexually connotative mask, like England', 'America' China' which enables it to engage with other national memes, make trade to aggregate its wealth or engage in war to aggregate its wealth.

This is entirely similar to what goes on in the stellar reals where with greater power the yang object may confront either other yang objects to either join together with them or consume them, or confront the yin object of stellar power either to devour it and destroy it to turn away from it to a safer haven in the universe than the place it had previously been in.

We have seen that in the relationship between macro and micro universes in the solar system where two types of macro systems reign and compete for hegemony the yang [supernova, suns etc.] and yin [black holes, neutron stars etc.] the microscopic content of these macroscopic entities will sacrifice their sexual identities [proton, neutron, electron order to aggregate the power needed to enable the macroscopic persona to better function.in other words protons, neutron, electron will change and interchange one another to achieve a total mass energy increase in the macroscopic entity in which they are located.

In the macroscopic world of stellar space however their intrinsic sexual identities on the macro scopes, sun vs black hole etc. The religiously jeep and it is in these clear personas of yin and yang that the battles, powers and time space organization of the universe is determined and parceled out.

To next phase of universal organization and differentiation is both more minuscule and more complex in the solar system of them across of stellar space trail of differentials follows the sun, father often solar system we can find modeled in the power system of brook Taylor, an era.

The combination of the yin with his offspring the 9 planets of our solar system we can find modeled in the power system of brook Taylor an English mathematician of the17th century. While aeno if and Archimedes of believed a system which skimmed infinitesimal quantities that converged to a finite sun existed this formula had to wait for Leibniz and newton to square the circle in the 17th century to be verified by Taylor

Taylor proposed in the formula that in a homogeneous linear function the sums of succeeded powers as succeeding differentials at the same point could be summed to <1. The prime integral to be thus differentiated can be seen to be the sun while succeeding powers, differentials at same point in the linear equating can be seen to be the sequence of planets of the sun from mercury, the first planet to Pluto the 9th.

$$f(a) + \frac{f'(a)}{1!}(x-a) + \frac{f''(a)}{2!}(x-a)^2 + \frac{f'''(a)}{3!}(x-a)^3 + \cdots.$$

$$\sum_{n=0}^{\infty} \frac{f^{(n)}(a)}{n!}(x-a)^n$$

However in the astronomy of our solar system there are strings of higher differentials in the moons of Jupiter and Saturn. These differentials are second order differentials have three variables instead of 2 and because their coefficients are no recurring or unknowable the general equation for second degree differentials will be non-homogeneous or even complex

Homogeneous Equations: If g(t)=0, then y″+p(t)y′+q(t)y =g(t).

Then we think of what we do know of these planets moons, this is understandable. Among the several moons of these plants that have life sustaining oxygen and oceans, on euro moon of Jupiter the oceans Lau miles beneath the surface of the planet while access to oceans are still unknown. Also as deep oceans are often buried below the planet surface and may lack access to the ocean streams of the surface the base composition of any life forms on the moons may be memes of our own and difficult to communicate with in any intelligible manner,These meme like life forms pay present in a variety of bizarre ways, including as tricksters or pranksters. Their mathematical configuration will be accretion of whole cardinal bases yielding holly imaginary numbers as bases with exponents of magnetic pols.

What kind of life form could we expect to find on the Europa moon?

Overall possible human like with extension provided by human like lungs, but body portioned between ocean swathes leaking out to other oxthenated swaths. Living partly in time present, partly millions of year time past. This life form would be bizarre and a real challenge to communicate with for humans.

On the other hand as the possible life form from Europe has several levels of unreality in him,he ould present as a trickster, and a trickster from our own time frame continuum

These bizarrely ineffective atomic elements modeling bizarrely ineffective planets we may call memes. In a sine the differentials may all be called memes of various low effectiveness as they are like scraps of the whole or even scraps of scraps of the whole. Just as the planet scan be sent from the orders of first degree differential following integer of 1, so they can be seen from descriptive mapping. In the descriptive mapping where we see the relative efficiency of planetary function following the model of full efficacy according the geometric outcome of the cardinal number. Atomic number designating them in the periodic table of the elements. Thus mercury atomic/cardinal number/map number 1 is synchronistic ally merged with hydrogen, atomic number 1, the element correlate of the sun itself, atomic 2/cardinal2, Venus mapped is so highly unstable it resembles quality of solar system life on mercury, highly volatile and unstable as its atomic element correlate helium.

Third rock from the sun, carousing the number 3, the banner of unreality and irrelevance to activities of macro scopes in greater stellar universe.

On planet no 4, modeling the atomic element 4, beryllium, a rare and minor element, planet no 4 has little of the vital elements needed for oxygenated biomass life

Planet no 5, modeling atomic element/cardinal number 5, also rare moderately unstable element, models plant Jupiter with vast qualities of oxygen, methane and co2, with none of the plant life sustainable for extension life, The planet no 5 is home to hundreds of moons, some of which promise bizarre life forms, in keeping with its similitude with mathematical model of bizarre second degree differential equation.

Planet no 6, unlike its atomic element model, is too cold to sustain the for a of a carbon rich environment unlike its predecessor Jupiter is awash with raw primal gasses of oxygen, methane, c02 and ice and like Jupiter Saturn is host to hundreds of moons whose composition has not yet been recorded Uranus, the 7th planet from the sun, is ice planet, also has vast quantities of the methane and c02 gasses. Neptune and Pluto on outer rim of our solar system are mostly rock and metal, although vast quantities of ice may also be present.

While in using brook Taylor's power series formula we are in visioning our line of planets stretching from mercury,the sun's first planet to Pluto on the edge of our solar system, through Venus, earth, mars, Jupiter, Saturn, Uranus,

Neptune on a way. While the line from the sun to Pluto i is aligned to be a straight line in curve euclidean space, where Jupiter and then Saturn intersect with this line, at 45 degree angle two other lines jut out on 90 degree angle holding in their target to ties hundreds of moons of various life sustaining resources.through these lines also pass the largest magnetic field in our solar system, of u to 5 magnetic poles. This magnetic field is a source of bizarre sound and radio waves from our galaxy hundreds of million light years away, does not radiate or receive light radiation on hospitable to life as we know it, is a deathly cold planet, a case may be made for it as black hole or black hole in the making.

The so called black body radiation considered characteristic of black holes cam be seen here to ne black body radiation of sound and radio waves rather than light or electromagnetic waves, which is more in keeping withe basic nature of black holes--as sound wave receptor--in thee first pace

What if this region of Jupiter's huge moving our solar system, the subject of so much religious speculation in the ancient kingdoms of Babylon, Egypt, China and Greece where the voice of god or supernatural beings was so often claimed to being heard one planetary system away from the rocky planets of earth and mars and a hop skip from terrestrial planets cherished superman belt of hundreds of thousands of tiny asteroids and comets, is really an intersection between light and dark matter in our solar system. between yin and yang in our backyard?

Take another look at the trajectories that extend out in 45 degree angles from Jupiter and Saturn-if the line from mercury to Pluto is really the trajectory of the fist differentials of our sun from mercury through Pluto then the trajectories that extend out from them are not the first differentials but the second? or what if both trajectories from sun to Pluto and from Jupiter.Saturn to their moons were equally probable both first and second differential. one trajectory emending from light force yang of our sun and other from much sought after yin force black hole, our dark sun of Jupiter?

In fact what we are see3ing as we closely examine all information pertaining to 5th planet from sun and beyond that all the remaining planets are trained but a prolific line of moons from in the hundreds to many thousands where these trajectories are at another 90 degree angle to the trail of planets from mercury to Pluto following our sun as first order differentials, Them m,any lines of mo0on following Jupiter and planets beyond such as Saturn, Uranus, Neptune and Pluto, including asteroid belt between mars and Jupiter and super belt of tiny asteroids after Neptune and Pluto consist of a gigantic plane. which is under the effect of the dark sun, Jupiter and are planetary bodies of dark matter, and are all second order differential.

This plane of dark matter intersects at 90 degree angles with the great plane of light matter emanating from the sun's first single differential mercury, all their way to Pluto, it can be seen that all planets in the single differential line from Jupiter to Pluto are interesting with the dark matter line of double differentials from Jupiter to Pluto and are in the area of the cross between yin and yang.the fact that the line of double differential is not a line per Se but a plane, a plane of dark matter what is a line projecting a plane of single differentials or light matter, means this section of intersection's geometric dimensions are not just third dimensional but fourth dimensional and we have achieved the **Minkowsky** space time where time, like space, goes not just forward, but also backward.

Furthermore corroborating this theorem is recent findings on the 'dwarf' planet Pluto dorm evidence of new horizons satellite, that Pluto's polar axis is a wandering polar, with shifting solar dipole in its 248 year revolution around the sun [see Nimmo, Keane, Binzel et all in NATURE, Nov 16, 2016]

We have calculated in previous chapters how quantum skein of our universe is 2 to the power of 6 for at quantum number since our universe, is a total of 64 different outcomes with base of 2 possible valences plus and minus to power of 6 where 6 is the number of possible units in any given code for possible outcomes and the 2 or 4 squares are possible magnetic poles of however we are at an intersection be ten dark and light matter, yin and yang, at the Jupiter junction we cross as it were through the looking glass, change charge and parity and we get a mirror image of the skein of quantity in numbers in the light matter or yang universe. instead of bas2 to power of 6 [on number if unit code] or the light universe with 64outvomes we get base6 to power of2 [or4] for or 36 or 72 outcomes/

In corroborating the 4th dimension of the space between the line of planets meaning from [forward time moving] yang line from Mercury to Pluto intersecting at right angles with [backward time in moving] yin plane [bounded but line projections of all planets' moons from Jupiter to Pluto] the finding that solar pole of Pluto is a polar wanderer is critical.

The wandering between time forward, yang, and backward, yin, is what makes its solar axis a 'wanderer'. Remember in all this, yin ' is the pervade of both life and death, the beginning and the end while yang is the living years between these two poles,'

The basis of this theorem should provide the long sought handle on the niggling problem of time travel in Western science.

As the base is of composed of 6 possible condone the base is an aggregate arithmetical result, imaginary numbers with 2 possibilities. we are strictly in he realm of imagination, imaginary phenomena for maybe children rather than adults-at any rate Lewis Carole is right around the corner as we fall into Alice's world in wonderland

Yet we have learned in our floras in yin and yang than no matter how rational our best rational minds tells us we are behaving, we are performing no actions nor thinking any thought considered rational according to behavior of yin and yang in the stellar universe. of dark matter universe only offers us some stellar slapstick compared to limber drama of our light matter universe perhaps it behaves us to be tolerant of friends unless fortunate earthly dwellings than our own"

Among the most striking differences between the macro of the greater universe and solar system life is that yin/yang only contact in the greater universe in a life or death struggle for survival while in solar system life yin and yang contact in love and friendship, bonding together in carbon compounds and oxides, cohabiting and reproducing. When yin [odd valence, hydrogen, nitrogen, phosphorous] bond with yang [H2, carbon, oxygen]. We gey igneous rocks, rocky mantles, plant life, animal life, water, male and females animal and plant genders. Nowhere in the solar system is entropy in greater supply and with greater attempts to see it preserved than dissipate to ashes in empty space.

From rocket laboratories to symphonic orchestras, from schools, hospitals, shopping malls, the 4 billion plus of human yin tangs all build their own entropic universes, whether as families, clubs, communities, nations, regions, and humans currently seek better and more effective ways of preserving entropy and making more of it rather than losing it or destroying it. Efforts to conserve entropy include schools, hospitals, peace seeking governments and nations and United Nations. Surrogates preserving entropy abound from schools to fight ignorance, medicine and hospitals to fight illness and death and preserve life, national and international movement it's of every kind to spread information and promote peace, choosing surrogates for raw sex in everything from same sex lovers to adored pets and beloved personal machines, deified clothing and movie and athletic stars, computers and cars. In the solar system universe humans carry the fight to model the works of the great creations of the universe, at a cheaper price, with wider distribution, more personalized effect rather than let it be dissipated and disappear in illness, death, war, personal, inter relational, familial, national conflict. The struggle goes on

APPENDIX TO CHAPTER 9, STELLAR SLAPSTICK   by Eleanor Morris Wu

Infinity-->0yin

A woman's life receding backward in control of the major life events of birth and death, yin or yang, proceeds of the original universe

in human life involves naming leading to conflict or non naming [tao] lack of naming+no conflict yin+ yang 0<1<infinity

IDENTITY where 1=person, always greater than zero, less than infinity, this is Confucius' Middle way, only a possibility in human life of post ancient universe in the solar system

CHAPTER 10
FRED HOYLE'S ASTROPHYSICS OF THE 21SR CEN-TURY: INCONVERTIBLE SCIENTIFIC ARGUMENT THAT INTELLIGENT DESIGN IS THE ORGANIZ1ING PRINCIPLE OF THE UNIVERSE

After ww2 as understanding of the atomic nucleus and nuclear science deepened, scientists attempted to use byproduct of nuclear science, spectroscopy, to understand the origin of elements of earth and in the universe, where they came from, where they made and what was their perceiver co, position on earth and in the universe. Strong supporters of Big Bang theory like George Lemaitre and George Gamow assumed the vast majority of chemical elements as described and listed in the Periodic Table of the Elements were produced in the Big Bang which had been determined to have occurred 13.8million years ago, However careful scientific examination had determined this could not be true as after 20 seconds the Big Bang had cooled to the point where it could no longer gus the protons and neutrons to make new chemical elements with higher atom numbers of 4, boron or 5, beryllium. The nucleo genesis furnace of the Big Bang was responsible for elements with atomic numbers from 1-5, and mostly the atomic elements of 1 hydrogen, 2 helium, and 3 lithium. As boron and beryllium had low enough atomic numbers to products of the Big Bang nucleo genesis furnace, their atomic composition was so unstable they usually quickly degenerated to lithium, hydrogen the most stable of the atomic elements produce in Big Bang nucleo genesis furnace.

Even though Big Bang nucleo genesis furnace only produce a few of the atomic elements known, those elements it did produce/ provided the most massive volume of atomic elements known-between them hydrogen and helium contain 98% percent by volume of all chemical elements in the universe.

Eccentric Cambridge University theoretical astrophysicist Fred Hoyle conceptualized that metalicity [chemical elements of higher=er atomic numbers] must be being produced in a secondary process in hot suns more than 8 times hotter then our sin]in stellar regions of deep spacer in super nova where massive suns were exploding and spilling their waste in huge gas clouds into entered tellar regions where these gas clouds might themselves coagulate and become new suns.

As a lifetime critic of the Big Bang THEORY [which he ironically had sarcastically named on a British talk show.

Hoyle was a proponent of the Steady State Theory, in which mass and energy did not have a dingle point origin as in the Big Bang but multiple points of origin , Hoyle envisioned space itself as part of matter and call space the viral C section, a composition of space-matter and negative change responsible for conservation of energy, In his conceptualization of the origins of the universe he envisioned space-matter constantly being created and never dying eternally rejuvenating himself. It was within this vision that he was able to conceptualize the 'secondary systems' in the universe that produced netukicuty, all elements of higher number than 5, in a universal self creating manifestation of hot suns burning hotter than 8times our own sun and providing the stellar furnaces in which all chemical elements with higher number of 5 were produced, to populate the universe and permit such wonders as self suntanning planetary systems of which earth and her fellow planets here were likely not the not the only explain the universe.

Although most of the work leading Hoyle to this way of thinking were done in the'40's after ww2, his most seminal --and most largely disregarded paper was written for Astronomy Journalism 1954. In this paper he described the structure and function of stars burning 6 times hotter then our sun in great detail and in using actual such subs as objects of investigation the spectral details which would turn all astrophysics up to that time into turmoil.

Hoyle envisioned these stars 8-10 times larger than our own sun and withe characteristic iron-nickel hot core burning itself out in time in discrete layers or shells until the entire sun was burnt out to a white dwarf,its effluences from the various layers or shells dispelled from the star as gases of precious metalicity to enrich stelar space for planetary and life potential.

The most critical of Hoyle's vision was the discreteness of the shells within the furnace sun which made this sun appear onion like.for it ensured that once an element was forged in the furnace sun, its purity would not be impinged upon by the element in the next or previous layer or shell of creation, It can be seen from the graph in Hoyle's 1954 paper how shell of the highest atomic number, mickel was created first followed by layers of elements of lower atomic numbers up to oxygen, carbon, the in the outermost shell elements of lowest atomic number, hydrogen and helium. When the sun burnt out discrete waves of intact elements were dispelled from the dying sun, to installer space where they populated planetary objects such as earth with their goodness.on earth as know from research done after Hoyle's time the crust, mantle and atmosphere are composed of the elements from carbon to nickle.

Furthermore Hoyle's unique induction, later confirmed by the experimental physicists at Cal State Institute of Technology, that the carbon had the unique resonance of 2,75 meV enabled the profusion of carbon to be produced by the furnace sun that would make carbon based life on earth possible,was another vital clue in the picture of intelligent design of the universe.

While there is no way for us to prove that all these life favoring coincidences were ordered by some Higher Being, or even that the concatenation of such life favoring series of coincidences were the product of some innate logical chain with possible casual connections, it cannot be denied by even the most atheistic denier of higher causes, that there was a definite life favoring bias in the concatenation of so many lied favoring coincidences and circumstances.

In defiance of all probabilistic indicators of non life favoring instances I think it can be said that the scientific basis for arguing intelligent design of the universe is incontrovertible.

While Hoyle did not spell out a formula fo the special life favoring elements from carbon to nickel, Hoyle's collaborator and follower, Donald Clayton felt the language in Hoyle's 1954 paper describing a formula was detailed enough for him to fill in the consequential algebra

Clayton;s reconstruction of Hoyle's famous equation was as follows

$dm(\text{C-Ni})/dt = BM>(t)$ **Evnucl** $\sum k\Delta mk$

where **Evnucl** expresses the nuclear and stellar evolution of a massive star, and $\sum k\Delta mk$ is the sum over k isotope masses.while B is the birth of their massive star DM at time t. The insertion of k as isotope of chemicals carbon to nickel reinforce the discreteness of the chemical elements produced in the furnace sun.

Here the massive star furnace M in shell segments delta mass carbon-nickel discrete elements in time since birth of star,The implication here that carbon to nickel was a special set could not have been known to Hoyle in his lifetime except through his incredible intuition. We know now they are a set as they have carefully been observed by experimental scientists to be the constituents of all aspects of earth from physical to organic chemistry level.

In a later 1957paper written in collaboration with WillyFowler, Geoffrey and Margaret Burbidge a full exploration of the rich spectroscopy product showing production of metalicity from carbon to nickel in the furnace sun is a landmark paper of stellar physics published in Reviews of Modern Physics in 1957.

Even though Fowler and the Burbidges won the Nobel Prize for this paper [The formal title of the paper is *Synthesis of the Elements in Stars*, but the article is generally referred to only as "B^2FH"--Hoyle had been wrongly eased out of the prize due to his failure to insert his formula into the paper], the full implications of the furnace sun discovery as a secondary avenue of element production remain lost to the scientific community.

It is my thesis that only with the opening of an entirely new scientific field, 'relative abundance of the elements' in the last part of the 20th century, that the full implications of Hoyle's spectroscopic work and his conceptualization of it can be fully appreciated.

In the periodic Table of the Elements, revised and revamped but Professor Sheehan of the UC Santa Cruz, according to the 'relative abundance of the elements' can we see in graphic form that it is neither te atomic number or position of an element in the Table that determines its significance but its numerical value determining the relative size of its actual occurrence vis a vis all other elements in the universe.

Hence we may rename Mendeleyev's Periodic Table of the Elements to 'Existential Table of the Elements [viewed i a Periodic Tableaux]'

Thus the significance of any atomic element is its actual range of existence within the table of all other atomic elements in the universe.

Since observational research has shown that the spectroscopic results of the furnace suns of elements from carbon to nickel are te elements most favorable to life and planetary life in the universe, revamping of the old Periodic Table shows it is these elements as well which have the highest incidence of existential occurrence,

Thus those elements most favorable to life on every level have the greatest existential existence in the universe, supporting our position that intelligent design--the modus operendi of favor-ability to life--is the organizing principle of the universe.

As the raw number array of the old Periodic Table has been shown by Professor Sheehan to be mathematically meaningless either as a template to predict structural or functional attributes of the atomic elements or to predict statistical probabilities for them, I propose moving the mathematical modeling of the Table of elements from atomic number [numerical position of element in the Table] to characteristic of the number itself determining the elements position in the array. I propose returning to serious analysis of the cardinal numbers in relation to atomic structure of element they designate, For significant elements with dual decimal I propose plotting the influence of the second digit upon te first or its cardinal number component.

I believe this kind of raw number analysis as I have used for cardinal numbers/atomic elements 1/hydrogen, 6/carbon, 7/nitrogen, 8/oxygen, 9/phosphorus in earlier sections of STELLAR SLAPSTICK will elucidate some of the structural and functional characteristics of significant existential elements of dual digits.

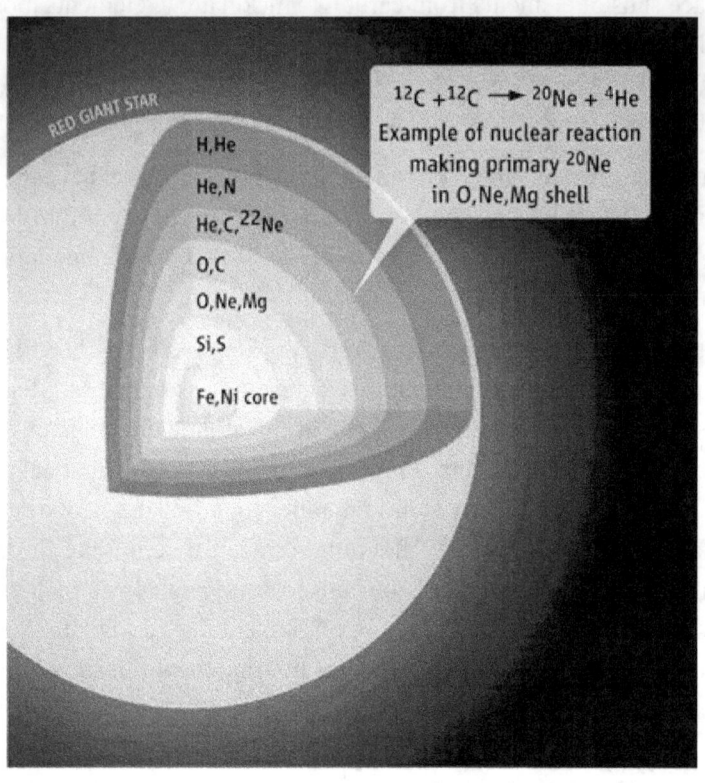

The Elements According to Relative Abundance

A Periodic Chart by Prof. Wm. F. Sheehan. University of Santa Clara, CA 95053
Ref. Chemistry, Vol. 49, No. 3. p 17-18. 1976

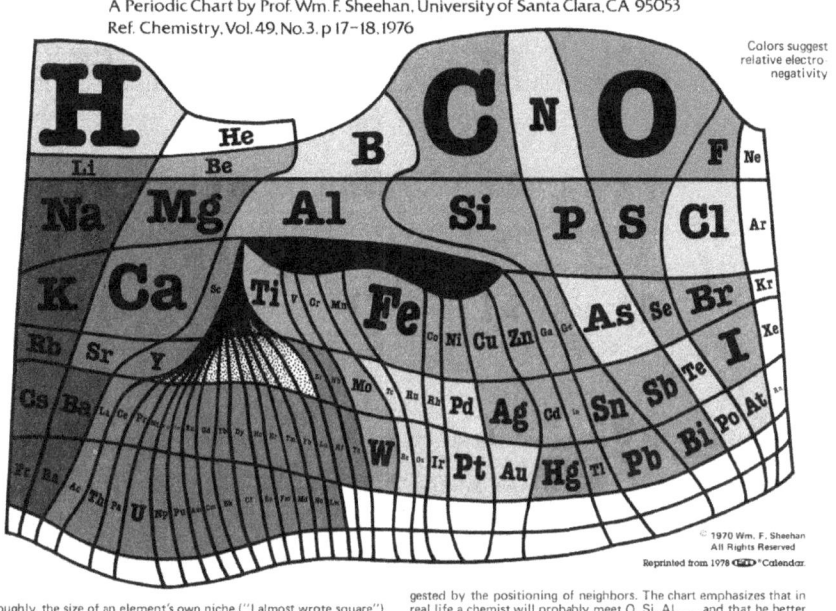

Roughly, the size of an element's own niche ("I almost wrote square") is proportioned to its abundance on Earth's surface, and in addition, certain chemical similarities (e.g., Be and Al, or B and Si) are suggested by the positioning of neighbors. The chart emphasizes that in real life a chemist will probably meet O, Si, Al, . . . and that he better do something about it. Periodic tables based upon elemental abundance would, of course, vary from planet to planet. . . W.F.S.

NOTE: TO ACCOMMODATE ALL ELEMENTS SOME DISTORTIONS WERE NECESSARY, FOR EXAMPLE SOME ELEMENTS DO NOT OCCUR NATURALLY

While the 1957 paper ignored the equation Hoyle had outlined for stellar nucleosynthesis being responsible for the abundance of elements between carbon and nickel, all elements responsible for carbon life and planetary mantle, crust, core and atmosphere in the solar system, as well as of course organic life, This paper was important, though few realized it at the time in recognizing stellar nucleo genesis, itself a secondary and improbablistic method of element creation, to the primary system of big bang nucleo genesis, It established that in the arcane features of stellar nucleo genesis that the abundance of elements from carbon to nickel that are responsible for all life forms in the universe and their critical subsects such as planetary structures, that will find intimation of intelligent design.

The 1954 paper comprehensively outlined the specific features of nucleo genesis and bed was left ou of the 1957 paper and was responsible for Hoyle being denied the Nobel Prize in , it instead being shared bt his collaborators Fowler and the Burbidges according to ban other of Hoyle's collaborators and devoted follower Clayton.

While in his 1954 paper Hoyle did not insert the critical algebra for te equation for stellar nucleo genesis, the descriptive language for the process was clear enough for Clayton to spell it out which he did in the paper he wrote about it in science in Clayton's reconstruction of Hoyle's famous equation was as follows

$dm(\text{C-Ni})/dt = BM{>}(t)$ **Evnucl** $\sum k\Delta mk$

Where **Evnucl** expresses the nuclear and stellar evolution of a massive star, and $\sum k\Delta mk$ is the sum over k isotope masses.while B is the birth of the massive star DM at time. The insertion of k as isotope of chemicals carbon to nickel reinforce the discreteness of the chemical elements produced in the furnace sun.

As spectropsy showed these massive stars produced metalicity from carbon to nickel. Hoyle;s description of the structure of these metalicity producing stars from the time of their prime to their dying days when they burned themselves out, spreading their precious meticility products into winds of stellar dust was of giant onion like spheres, Like onions each layer of the sphere burned to its peak creating the metalcity that that temperature decreed, then started burning on the next layer etc so each new layer produced in the metalcity of its previous layers ashes, starting with the burning of the nucellar core through to layers from nitrogen to oxygen, carbon and then to layers of the lightest elements oof helium and hydrogen.

One of the most remarkable results of Hoyle's tracing production of the elements or metalicity through its primary production units, the stellar furnaces of suns ten time shorter than our sun, is that these elements produced are not keyed to the template of the periodic table or bare array number system but a template of the relative abundance of the elements in the universe.

The chart prepared by Prof D. Sheehan of UC Santa Cruz shows beyond a shadow of a doubt that metalicity geared to the raw number array of the periodic table with its associate area of elements is statistically meaningless. Only a template that combines the coordinates of relative abundance of elements swith raw number array of the periodic table could provide a template meaningful enough to make valid predictions on the significance of various metalicities, These would be significance geared to te interests of life bearing beings, suggesting intelligent design in universal creation. Other suggestions of intelligent design is the profusion of life giving elements,carbon oxygen-,hydrogen, nitrogen and phosphor in the metalicity matrix. Further Hoyle's induction later proved true by experimental physics that carbon in the furnace of the massive sun 10 times the size of our own, possessed the special resonance of 2,7meV allowing the production of them asses amounts of carbon that is the prerequisite of all life, to be dispersed and spread throughout interstellar space to land in the vital regions where life became possible such as in our solar system.

In my next book I shall be working on creating the template of metalicity which makes Judgments of significances for life possible,using the unique power of the numbering systems of cardinal numbers with their consequent ramifications in atomic structures of these cardinal numbers and in dual digital numbers the unique virility in the combinations of cardinal numbers and their consequences in both physical and organic chemistry.